Carbon Strategies

Also by ANDREW HOFFMAN

Organizations, Policy and the Natural Environment: Institutional and Strategic Perspectives, editor with Marc Ventresca

From Heresy to Dogma: An Institutional History of Corporate Environmentalism—Expanded Edition

Competitive Environmental Strategy: A Guide to the Changing Business Landscape

Global Climate Change: A Senior Level Dialogue, editor

Carbon Strategies

How Leading Companies Are Reducing Their Climate Change Footprint

Andrew J. Hoffman

With contributions from:
DOUGLAS GLANCY
MICHAEL HORN
SCOTT PRYOR
MARK SHAHINIAN
GREGORY SHOPOFF

Ann Arbor
The University of Michigan Press

Published in the United States of America by
The University of Michigan Press
Manufactured in the United States of America
∞ Printed on acid-free paper

2010 2009 2008 2007 4 3 2 1

A CIP catalog record for this book is available from the British Library.

U.S. CIP data applied for.

ISBN-13: 978-0-472-03265-5

ISBN-10: 0-472-03265-8

Contents

Foreword *vii*

Acknowledgments *ix*

Executive Summary *x*

Synthesis Report

Introduction *1*

A. The Business Case for Climate Action Grows *2*
B. Scope *5*
C. Methodology *5*
D. Overview *6*
E. Over-Arching Themes *6*

Stage I: Develop a Climate Strategy *10*

Step 1. Conduct an Emissions Profile Assessment *10*
A. Lessons Learned *10*
B. Emission Types *10*
C. Emission Metrics *12*
D. Emission Measurement Tools and Techniques *12*

Step 2. Gauge Risks and Opportunities *14*
A. Lessons Learned *14*
B. Benchmarking *14*
C. Risks from Operations, Products, and Service Lines *15*
D. Product and Service-Line Opportunities *15*

Step 3. Evaluate Options for Technological Solutions *18*
A. Lessons Learned *18*
B. Low-Hanging Fruit *18*
C. "Silver Bullets" *19*
D. Ongoing Reductions *20*
E. On-System versus Off-System Reductions *20*

Step 4. Set Goals and Targets *21*
A. Lessons Learned *21*
B. Motivating Factors *22*
C. Developing Climate Goals and Targets *24*
D. Differentiating GHG-Reduction and Energy-Efficiency Targets *27*
E. Making the Business Case for Climate Strategies *28*
F. Other Related Climate Goals and Targets *30*
G. Adaptation Strategies *31*

Stage II: Focus Inward 33

Step 5. Develop Financial Mechanisms to Support Climate Programs 33
A. Lessons Learned 33
B. Cost Estimates for GHG Reductions 33
C. Internal Carbon Trading 35
D. External Carbon Trading 35
E. Other Financial Instruments 36

Step 6. Engage the Organization 37
A. Lessons Learned 37
B. Gaining Buy-In 38
C. Senior Leadership 39
D. From Idea to Adoption 41
E. Moving Climate Change from the Periphery to the Core 42

Stage III: Focus Outward 48

Step 7. Formulate a Policy Strategy 48
A. Lessons Learned 48
B. The Link between Policy and Strategy 48
C. Policy is on the Horizon 49
D. Options for Policy Mechanisms 50

Step 8. Manage External Relations 53
A. Lessons Learned 53
B. Target Audience 53
C. External Resistance 58
D. Supply-Chain Partnerships 60

Conclusions 63

Case Studies 65
Cinergy: Managing "Stroke of the Pen" Risk 66
Swiss Re: Staying One Step Ahead on Climate Change, Not Two 78
DuPont: Shifting from Risk Management to Business Opportunity 90
Alcoa: Weaving Climate Change into the Business Case 103
The Shell Group: Maintaining a Seat at the Table 113
Whirlpool: Don't Switch Tracks When the Train Is Already Moving 123

Appendix A. A Compendium of Climate-Related Initiatives Used by BELC Companies 131

Appendix B. Glossary 165

Notes 167

Foreword *Eileen Claussen, President, Pew Center on Global Climate Change*

There is a growing consensus among corporate leaders that taking action on climate change is a responsible business decision. From market shifts to regulatory constraints, climate change poses real risks and opportunities that companies must begin planning for today or risk losing ground to their more forward-thinking competitors. Prudent steps taken now to address climate change can improve a company's competitive position relative to its peers and earn it a seat at the table to influence climate policy. With more and more action at the state level and increasing scientific clarity, it is time for businesses to craft corporate strategies that address climate change.

In this book, author Andrew Hoffman of the University of Michigan, with the assistance of five graduate students, has developed a "how to" manual for companies interested in developing effective climate strategies. One of the clearest conclusions is that businesses need to engage actively with government in the development of climate policy. The U.S. Climate Action Partnership (USCAP), a coalition of major companies and leading non-governmental organizations, including the Pew Center, offers perhaps the best example of increased corporate involvement in the climate policy debate. In January of 2007, USCAP released a landmark set of principles and recommendations to underscore the urgent need for a policy framework on climate change. The group, which includes Alcoa, GE, Caterpillar and DuPont, recommends a cap-and-trade system as the cornerstone of a national climate policy, and will push lawmakers to enact legislation to reduce greenhouse gas emissions at the soonest possible date. Other companies and trade associations are also becoming more involved. The Edison Electric Institute, the leading trade association for electric utilities, has released a set of principles to guide its engagement in the climate change debate.

These companies recognize that after years of inaction at the federal level, momentum is growing to pass mandatory climate legislation. Nearly all the companies surveyed for this book believe that federal legislation is imminent, and 84 percent of those believe federal standards will take effect before 2015. With a number of new climate bills introduced, it is clear that Congress has entered the design phase for legislation. Now is the ideal time for the corporate sector to engage constructively with lawmakers to ensure that sensible policy is developed to reduce greenhouse gas emissions at the lowest possible cost.

Furthermore, constructive engagement is tightly linked with another compelling theme of this book: the shift of companies' focus to creating climate-related market opportunities. Companies with a strong history of reducing emissions are shifting their focus from risk management to exploring new business platforms. They understand better than their peers that new markets will be created and existing ones will change. There will be winners and losers. The shape of climate legislation will be the strongest factor in determining how the market rewards innovators of climate-friendly products and services, as well as how it punishes laggards. More than ever, integrating climate issues into corporate strategy is a necessary aspect of managing risk and seizing competitive advantage.

The Pew Center would like to thank Mike Lenox, Forest Reinhardt, and Paul Tebo for their comments on an earlier draft of this book; Alcoa, Cinergy (now Duke Energy), DuPont, the Shell Group, Swiss Re, and Whirlpool for agreeing to be profiled for the case studies; all the companies that completed the Corporate Strategies Survey; and the many member companies of our Business Environmental Leadership Council that provided comments and guidance throughout the research process.

Acknowledgments

I would like to thank the following people for assistance in developing the case studies: Pat Atkins, Ken Martchek, Richard Notte, Randy Overbey, Jake Siewert, and Vince Van Son (Alcoa); Eric Kuhn, Kevin Leahy, David Maltz, Darlene Radcliffe, Jim Rogers, Catherine Stempien, and John Stowell (Cinergy); John Carberry, Uma Chowdhry, John DeRuyter, Linda Fisher, Craig Heinrich, Don Johnson, Mack McFarland, Ed Mongan, Michael Parr, James Porter and Dawn Rittenhouse (DuPont); David Hone (Shell); Nigel Baker, David Bresch, Pascal Dudle, Ivo Menzinger, Andreas Schlaepfer, Cosette Simon, Brian Thomas, Chris Walker and Mark Way (Swiss Re); Tom Catania, Dick Conrad, Mark Dahmer, JB Hoyt, Bob Karwowski, Casey Tubman and Steve Willis (Whirlpool). I would also like to acknowledge the assistance of Truman Semans, John Woody, Judith Greenwald, Timothy Juliani, and Andre de Fontaine from the Pew Center on Global Climate Change in developing this book, with particular acknowledgement to Andre de Fontain, John Woody, and Truman Semans for authoring the climate-related initiatives in appendix A. Finally, I would like to thank the research team that contributed to the production of this work: Douglas Glancy, Michael Horn, Scott Pryor, Mark Shahinian, and Gregory Shopoff.

All greenhouse gas emissions generated from air and road travel associated with the development of this book have been offset through renewable energy certificates purchased from 3 Phases Energy Services, LLC. In total, 14 metric tons of carbon offsets were created through 18.9 MWh of landfill-gas electricity generated at the Granger Electric Generating Station in Grand Blanc, Michigan.

Andrew J. Hoffman
Holcim (U.S.) Professor of Sustainable Enterprise
The Frederick A. and Barbara M. Erb Institute for Global Sustainable Enterprise
The University of Michigan
Ann Arbor, Michigan

Executive Summary

This book compiles the experience and best practices of large corporations that have developed and implemented strategies to address climate change. Based on a 31-company survey, six in-depth case studies, a review of the literature, and experience gained by the Pew Center in working with companies in its Business Environmental Leadership Council (BELC), the book describes the development and implementation of climate-related strategies. It is primarily a "how to" manual for other companies interested in developing similar strategies. But the book will also be of value to investors and analysts in evaluating the effectiveness of company strategies for managing climate risk and capturing climate-related competitive advantage. Finally, it will offer policymakers insight into corporate views on greenhouse gas (GHG) regulation, government assistance for technology advancement, and other policy issues. Although the book focuses primarily on U.S.-based multinationals, it considers the global context of climate change and related market transformation.

The book describes eight specific steps clustered into three stages that describe the various components of a climate-related strategy. Table ES1 summarizes these steps, which include assessing emissions and exposure to climate-related risks, gauging risks and opportunities, evaluating action options, setting goals and targets, developing financial mechanisms, engaging the organization, formulating policy strategy, and managing external relationships. The book is organized along the framework presented in the table, though it must be emphasized that individual companies do not necessarily follow the steps shown in a linear fashion.

Lessons learned at each step of the strategy development process are presented throughout the book. Taken together, four overarching themes emerge from the survey results and case studies. The first is the importance of **strategic timing.** Some companies acknowledge the dangers of starting too early on climate action, while others highlight the risks of starting too late. Despite continuing uncertainty, there is general consensus among the companies in this book that recent changes in the level of external awareness about climate risks, state government action, momentum toward stronger federal policy, and consumer demand for cleaner and more efficient products make it imperative to act now. Well-timed strategies can prepare companies for eventual regulation and create flexibility for longer-range strategic options.

A second theme is the importance of **establishing an appropriate level of commitment.** While the companies in this book are leaders in their industries, some caution against getting too far ahead of the competition. For many companies, uncertain demands from government, the marketplace, and the financial community—coupled with limited hard data and models to guide aggressive action—make it challenging to support extensive expenditures on GHG reductions. Therefore, many companies justify early action on other grounds: the managerial imperative to undertake low-risk initiatives that produce immediate or near-term cost benefits; their fiduciary obligation

Table ES1

Stages of Climate-Related **Strategy Development**

Stage I Develop a Climate Strategy				Stage II Focus Inward		Stage III Focus Outward	
Assess Emissions Profile	**Gauge Risks and Opportunities**	**Evaluate Action Options**	**Set Goals and Targets**	**Develop Financial Mechanisms**	**Engage the Organization**	**Formulate Policy Strategy**	**Manage External Relations**
What kinds of direct and indirect GHG emissions are being created, from what sources, and in what quantities? What metrics can be used to track emissions, and what technologies or techniques are required to measure them?	What risks are posed by emissions from operations and GHG-intensity of products and services? Where can we excel and get ahead of peers in climate-friendly or risk-reducing business lines? How may demand for products and services change? What products and services may flourish given carbon constraints?	What options are available for reducing emissions? Are there any "low-hanging" emission-reduction opportunities? Where can we innovate? What long-run steps can be taken? How can climate-related strategies enhance top-line and bottom-line objectives?	Why set GHG reduction targets? What kinds of efficiency or reduction targets should be set, and over what time period? How do efficiency improvements relate to GHG reductions? How can targets be connected to business strategy? What kind of goals are achievable regarding new business opportunities? What kind of adaptation strategies should be considered?	What financial instruments are available to support GHG reductions? What are the pros and cons of emissions trading (internal and external), carbon shadow pricing, lower hurdle rates, and special capital reserves?	How can buy-in from the workforce be achieved? How important is senior leadership? Where are the sources of support and resistance? How can resistance be overcome? How can climate-related activities move from the periphery to the core?	How might possible policies help or hurt business and/or on-going climate-related activities? What policy options are on the table? What is a desirable policy outcome? What are the best ways to influence climate policy at the state, national, or international level?	What external constituents are important to the success of climate-related strategies? How should they be engaged?
Step 1	**Step 2**	**Step 3**	**Step 4**	**Step 5**	**Step 6**	**Step 7**	**Step 8**

Feedback and monitoring to refine business case, strategy elements, and tactics to address risks from climate change and from related regulations, particularly to the extent these could affect future asset values and market positioning; and socially and ethically responsible business values—that is "doing the right thing."

A third theme for many companies is the need to **influence policy development**. Any policy that regulates GHG emissions will certainly constitute a major market shift, setting new "rules of the game" and changing the competitive landscape. Companies in this book feel they cannot leave the ultimate form of such regulations to chance. All policies are not equal; they will, by their nature, favor certain actions, companies, and industries. Early action is seen as a way for companies to gain credibility and leverage participation in the process of policy development, and thereby have a measure of control over their future business environment.

A fourth and final theme is the importance of **creating business opportunities.** Companies with a history of climate-related activity are trying to shift their strategies from a focus on risk management and bottom-line protection to instead emphasize business opportunities and top-line enhancements. Firms that incorporate climate change into their core business strategies will be in the best position to take advantage of emerging opportunities and gain competitive advantage in a changing market environment. Sustainable climate strategies cannot be an add-on to business as usual; they must be integrated with a company's core business activities.

In the end, it is the consensus of the companies in this book that climate change is driving a major transition—one that will both alter existing markets and create new ones. As in any such transition, there are risks and opportunities, and there will be winners and losers. In this context, a growing number of companies believe that inaction is no longer a viable option. All companies will be affected to varying degrees, and all have a managerial and fiduciary obligation at least to assess their business exposure to decide whether action is prudent.

Synthesis Report

Introduction

This book compiles the collected wisdom and experience of companies with a history of addressing climate change. It provides a model for corporate action, from making the business case for a proactive approach, to developing appropriate goals and targets, to implementing innovative strategies. Exploring the risks, rewards, opportunities, and barriers companies have encountered and documenting their successes and failures yields insights for those considering similar action and suggests best practices for assessing future efforts. As Yolanda Pagano, Director of Climate Strategy and Programs at Exelon, explains, "Many others—companies, governments and NGOs—have plowed this road before. Seek to leverage their learnings."

One prime motivation for early action on climate change is the looming threat of greenhouse gas (GHG) controls. Nearly all companies in this book (90 percent) believe that government regulation is imminent, and 67 percent believe it will take effect between 2010 and 2015 (see Figure 1). All face systemic risks from changing policies and higher energy and feedstock prices as a result of GHG controls. But they also have individual reasons for addressing the issue. Some companies are deeply engaged in the scientific debate over climate change. For others, that debate is secondary to the potential business impacts of regulation. Still others look to market changes and opportunities caused by shifting consumer and investor demands. All companies see a business reason for undertaking climate-related strategies, and each of their strategies reflects a different sense of the changing business and policy environment.

Nearly all companies in this book (90 percent) believe that government regulation is imminent, and 67 percent believe it will come between 2010 and 2015.

Figure 1

Anticipated Date of Federal Standards on Climate Change

[If you believe that federal standards on climate change are imminent] when do you believe these standards will take effect?

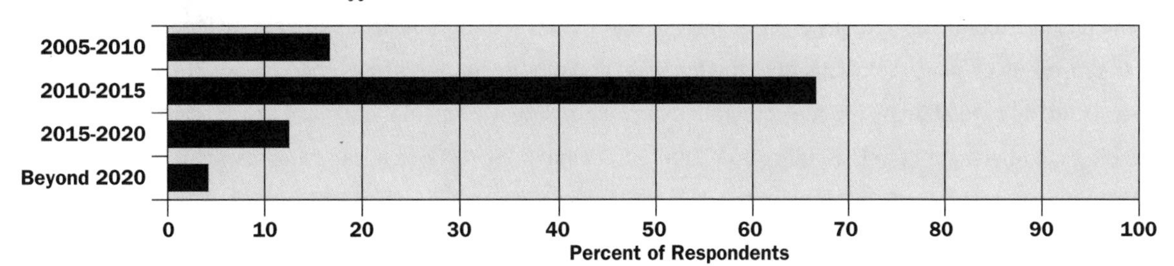

Total Respondents: 24

A. The Business Case for Climate Action Grows

Is a carbon-constrained world inevitable? Should companies engage in the policy debate? Are there business opportunities in this changing landscape? Increasingly, the answer to such questions is yes.[1]

Climate change creates systemic risks across the entire economy, affecting energy prices, national income, health, and agriculture, and creating regulatory, physical, and reputational risks at the sector, industry, and company-specific levels.[2] In short, climate change is altering the competitive environment, and certain companies, industries, and sectors will be more at risk than others. Some see the electric utility, steel, and aluminum industries as particularly vulnerable.[3] Others warn of impacts to oil and gas companies[4] or automakers.[5] Some see American companies overall as less prepared than their European and Asian counterparts.[6] Few sectors are immune from these effects.

In the public arena there are signs that a national climate policy is very near. As of June 2007, almost 600 mayors representing over 67 million Americans had signed the U.S. Mayors Climate Protection Agreement, which urges "the U.S. Congress to pass the bipartisan greenhouse gas reduction legislation, which would establish a national emissions trading system." Much like the process that led to the formation of the Environmental Protection Agency (EPA) in 1970,[7] states are increasingly enacting climate-related legislation. Nearly all states have now enacted some form of climate-related policy, and a growing number are joining regional greenhouse gas (GHG) reporting programs. Fifteen states have GHG emission reduction targets; 34 have climate action plans completed or in progress; 24, plus Washington, DC, have established renewable energy portfolio standards; and 34 have mandates or incentives promoting ethanol.

Activity in Congress has also been increasing, spurred by state action and several other influential trends and developments. In 2003, the Senate defeated the McCain-Lieberman GHG cap-and-trade bill by a narrow vote. In 2005, however, a majority of the Senate supported a nonbinding resolution sponsored by Senator Jeff Bingaman calling for a "national, mandatory, market-based program to slow, stop, and reverse growth of [GHG] emissions." The Senate Energy Committee held the first hearing on designing such a program in April 2006, and several major corporations testified.[8] The momentum in Congress increased further following the 2006 mid-term elections. In the House, shortly after taking over as Speaker, Nancy Pelosi declared that climate change will be a priority on her agenda.[9] The change in power also resulted in new leadership in all the key committees related to climate, and most of these new chairs strongly favor passing climate legislation.

Another key factor driving increased Congressional action on climate change is the proactive engagement by the business community. Many analysts believe that the most important factor in passing legislation is clear consensus on policy solutions from the business community. In that sense, the January 2007 call for federal climate regulation by the ten corporations and four nongovernmental organizations that made up the founding membership of the U.S. Climate Action Partnership (USCAP) was a watershed moment in the history of U.S. climate policy.[10] USCAP has publicly recommended to Congress an economy-wide U.S. climate policy including a cap-and-trade system, goals for U.S. global leadership, and mandatory measures and incentives to cut emissions in transportation, buildings, and coal-based energy. Although there have been important

NGO-business collaborations on environment and social issues in the past, this partnership has achieved an unprecedented combination of industry member diversity and size, detailed recommendations on legislation, and CEO-level outreach to Congress.[11] USCAP has since expanded and now includes 23 major corporations and six leading NGOs.

The USCAP announcement helped stimulate a wave of Congressional hearings on climate policy, many featuring USCAP members invited by Congress to testify on the initiative. As of July 2007, there had been over 90 climate-related hearings held by a range of Congressional committees, including: Senate Commerce, Science, and Transportation; Senate Environment and Public Works; Senate Energy and Natural Resources; Senate Finance; House Energy and Commerce; House Oversight and Government Reform; House Science and Technology; House Ways and Means; House Transportation and Infrastructure; House Natural Resources; House Agriculture; and House Foreign Affairs.

Additionally, there has been a clear uptick in the number of climate change bills introduced by members of Congress. In the first six months of the 110th Congress there had been 120 climate-related bills introduced, eclipsing the 105 introduced in the entire 109th Congress. Members who have introduced or are working on legislation include:

- Senators Joe Lieberman and John Warner, who are crafting a bill based largely on the USCAP recommendations.
- Senate Environment and Public Works Committee Chair Barbara Boxer and Senator Bernie Sanders, who have introduced a bill that would cap emissions at 1990 levels by 2020 and seek additional reductions beyond that.
- Senate Energy and Natural Resources Committee Chair Jeff Bingaman, whose discussion draft is based on recommendations made in 2004 by the National Commission on Energy Policy to set up a cap-and-trade system with mechanisms to control the costs of emission credits.
- Senators Dianne Feinstein and Tom Carper, who have introduced a bill that would cover only the electric utility sector, capping emissions at 1990 levels by 2020, followed by progressively steeper reductions.
- Senators John Kerry and Olympia Snowe, whose bill aims to reduce emissions 60 percent below 1990 levels by 2050.

Movement can also be observed in other arenas. On the financial front, mainstream investors are beginning to take notice[12] with companies like Goldman-Sachs, Bank of America, JP Morgan, Chase, and Citi adopting guidelines for lending and asset management aimed at promoting clean-energy technologies.[13] The proposed $45-billion buyout of TXU, a major Texas electric utility company, by two private equity groups provides compelling evidence that the financial world is taking seriously the risks from climate change. As part of the proposed buyout, the private equity firms have agreed to scale back plans to build close to a dozen coal-fired power plants, support federal climate protection legislation, reduce GHG emissions to 1990 levels by 2020, and support a $400 million energy efficiency program.[14]

The intersection of fiduciary responsibility and climate risk is also coming into focus, particularly around the "materiality" of GHG emissions under the Sarbanes-Oxley Act of 2002,[15] which some believe creates new climate-related legal risks for companies (and their directors). This possibility is not just hypothetical: eight states and New York City have filed a lawsuit against five of the nation's largest power companies demanding that they cut carbon dioxide (CO_2) emissions.[16] Some major insurers have since expressed concern about exposure to Directors' and Officers' (D&O) liabilities if climate risk is not properly disclosed and/or addressed, even as the number of shareholder resolutions requesting financial-risk disclosure and plans to reduce GHG emissions grew from 20 in 2004 to 30 in 2005.[17] The Carbon Trust forecasts that "climate change could become a mainstream consumer issue by 2010," placing corporate brands at risk.[18]

On the technology front, President Bush laid out new priorities for energy research in his 2006 State of the Union address. As coal is expected to figure prominently in future energy supplies, not only in the United States but worldwide, attention has focused on new, high-efficiency coal combustion options such as integrated gasification combined-cycle (IGCC) technology and carbon capture and sequestration, as well as on the next generation of nuclear technology.[19] Clean-energy markets are also growing dramatically: annual combined revenue for solar photovoltaics, wind power, biofuels, and fuel cells jumped nearly 39 percent in one year, from $40 billion in 2005 to $55 billion in 2006. Clean Edge, a cleantech market research and consulting firm, projects that these four technologies will continue to grow and could become a $226 billion market by 2016.[20] Announcing a set-aside of $100 million for investments in cleaner energy, transportation, air, and water technologies, venture capitalist John Doerr of Kleiner Perkins Caulfield & Byers has said, "This field of greentech could be the largest economic opportunity of the 21st century."[21] Rising energy prices have also affected all areas of the economy and have strengthened the business case for reducing energy consumption, while creating new demand for hybrid vehicles and efficient appliances.[22]

Americans have become more attuned to the potential consequences of climate change in the wake of recent natural disasters, as has the insurance industry, which faced $46 billion in losses from such catastrophes in 2004,[23] the largest amount to date.[24] Future warming could disproportionately affect vulnerable sectors such as agriculture, fisheries, forestry, health care, insurance, real estate, and tourism—as well as offshore energy infrastructure (such as oil rigs and pipelines)[25]—prompting many in those sectors to begin exploring adaptation strategies.[26] Meanwhile, the scientific community continues to develop research and data on a variety of possible impacts, including glacial melt, sea-level rise, ocean acidification, and associated impacts on global water currents.[27] Indeed in the mainstream scientific community, the issue is no longer whether climate change is happening, but what can be done to slow its progress and mitigate its effects.[28]

All of these developments create an increasingly compelling case for corporate action on climate change. Indeed, according to Ceres, the number of American companies addressing the issue has risen notably just since 2003.[29] In this changing business environment, action by one company can affect many others. Wal-Mart, for example, recently announced that it will begin giving preference to suppliers who set goals for aggressively reducing GHG emissions (see text box on page 45),[30] while Toyota has been able to take market share from other automakers by aggressively pursuing more energy-efficient vehicles.[31] In light of all the growing signs, the Conference Board warns, "Businesses that ignore the debate over climate change do so at their peril."[32] The magnitude of this peril is coming into greater clarity. A recent report by Sir Nicholas Stern, former chief economist for the World Bank, points out that the economic benefits of early action to curb greenhouse gases would far outweigh the costs.[33]

B. Scope

This book focuses on "climate-related strategies"—defined as the set of goals and implementation plans within a corporation that are intended to reduce GHG emissions, produce significant GHG-reduction co-benefits, or that otherwise respond to climate-related changes in markets, public policy, or the physical world. Corporate activities encompassed by this rubric include measures for achieving direct and indirect emission reductions from a company's own operations (such as energy efficiency initiatives); research, development, and investment in low-carbon production and process-related technologies as well as climate-related financial and business services; reductions obtained through emissions offsets and trading; activities to reduce "upstream" or "downstream" emissions along the value chain; and adaptation strategies.

C. Methodology

The research team for this book utilized two data-gathering methods. The first was a 100-question survey of 27 members of the Business Environmental Leadership Council (BELC) of the Pew Center on Global Climate Change[34] and four non-BELC members.[35] The survey sample was weighted toward large, publicly-held, multi-national corporations based in North America (see Table 1).

The second data collection method involved six in-depth case studies of five BELC member companies[36] and one non-BELC member,[37] each of which had a stated commitment to reduce GHG emissions. To develop the case studies, the research team conducted face-to-face and telephone interviews with key executives and managers, typically including vice presidents for environment, health and safety (EHS); sustainability managers; operations managers; research and development personnel; and senior managers in governmental affairs and communications. Interviewers raised a consistent set of questions and topics to assure comparability between case studies and augmented the data gathered, where relevant, with information from secondary literature. The Pew Center has gathered feedback from BELC companies throughout the process.

Table 1

Survey **Demographics**

Category	Results
Sector Representation	Electric Utility: 28 percent High Tech: 9 percent Metals and Mining: 9 percent Oil and Gas: 9 percent Other*: 46 percent
Ownership Status	Public: 87 percent Private: 13 percent
Headquarter Location	North America: 90 percent
Multinational Operations	Yes: 72 percent No: 28 percent
Market Segment**	Business-to-Business: 47 percent Business-to-Customer: 60 percent
Annual Revenue	< $1B: 10 percent $1-10B: 45 percent $10-100B: 45 percent

* Other includes the following: Chemicals, Consumer Goods, Pharmaceuticals, Paper and Forest Products, and Cement.

**This figure exceeds 100 percent because some companies offer both services.

D. Overview

This book has three parts. **Part One** synthesizes the main findings. It is organized into eight sections, each of which covers a major step in the development of climate-related strategies. Table 2 summarizes these steps, which include assessing emissions and exposure to climate-related risks, gauging risks and opportunities, evaluating action options, setting goals and targets, developing financial mechanisms, engaging the organization, formulating policy strategy, and managing external relationships. While presented in a linear fashion, it is important to note that the steps shown are not always followed sequentially since companies' actions must be tailored to their organizational culture, capabilities, and business plan. As with any well-managed initiative, constant monitoring and feedback are essential to effective implementation.

Part Two consists of six detailed case studies. Each seeks to identify what is unique about a company's approach, as well as what is transferable and potentially useful to other companies considering climate-related strategies.

Appendix A offers a compendium of climate-related initiatives used by BELC companies (as of February 2007). The appendix is organized into seven areas and provides a valuable list of projects: Energy Supply Solutions; Energy Demand Solutions; Process Improvements; Waste Management Solutions; Transportation Solutions; Carbon Sequestration and Offsets Solutions; and Emissions Trading, Joint Implementation (JI) and Clean Development Mechanism (CDM) Solutions.

Taken in its entirety, this book offers a comprehensive "how to" guide for implementing climate-related strategies and a compendium of best practices in the field. The results should be of interest to corporate decision-makers who are developing or considering climate-related strategies and to others seeking to understand how companies gain competitive advantage by preparing for climate constraints, including financial analysts, institutional investors, state and federal officials, non-governmental organizations (NGOs), scholars, and participants in international efforts to address climate change.

E. Over-Arching Themes

Four over-arching themes emerged from the survey and case studies. These themes cut across all elements of climate-related strategies and involve timing, commitment, policy development, and business opportunity.

Ensure strategic timing: The question for companies in this book is not *whether* to take action on climate change, but *when*. Some have acknowledged the danger of pursuing initiatives too early; a few executives specifically highlighted false starts with the Clean Development Mechanism,[38] which many believe is not realizing its full potential (see "Frustration with the Clean Development Mechanism" on page 99). In contrast, other companies wish they had started earlier. The key to a successful strategy lies in correctly timing its various components. According to Ron Meissen, Senior Director of Environment, Health and Safety Engineering at Baxter International, "Companies should take action now to define their global climate-related strategy, set GHG reduction goals and implement GHG reduction activities, not just for environmental reasons, but also for competitive advantage."

Table 2

Stages of Climate-Related **Strategy Development**

Stage I Develop a Climate Strategy				Stage II Focus Inward		Stage III Focus Outward	
Assess Emissions Profile	**Gauge Risks and Opportunities**	**Evaluate Action Options**	**Set Goals and Targets**	**Develop Financial Mechanisms**	**Engage the Organization**	**Formulate Policy Strategy**	**Manage External Relations**
What kinds of direct and indirect GHG emissions are being created, from what sources, and in what quantities? What metrics can be used to track emissions, and what technologies or techniques are required to measure them?	What risks are posed by emissions from operations and GHG-intensity of products and services? Where can we excel and get ahead of peers in climate-friendly or risk-reducing business lines? How may demand for products and services change? What products and services may flourish given carbon constraints?	What options are available for reducing emissions? Are there any "low-hanging" emission-reduction opportunities? Where can we innovate? What long-run steps can be taken? How can climate-related strategies enhance top-line and bottom-line objectives?	Why set GHG reduction targets? What kinds of efficiency or reduction targets should be set, and over what time period? How do efficiency improvements relate to GHG reductions? How can targets be connected to business strategy? What kind of goals are achievable regarding new business opportunities? What kind of adaptation strategies should be considered?	What financial instruments are available to support GHG reductions? What are the pros and cons of emissions trading (internal and external), carbon shadow pricing, lower hurdle rates, and special capital reserves?	How can buy-in from the workforce be achieved? How important is senior leadership? Where are the sources of support and resistance? How can resistance be overcome? How can climate-related activities move from the periphery to the core?	How might possible policies help or hurt business and/or on-going climate-related activities? What policy options are on the table? What is a desirable policy outcome? What are the best ways to influence climate policy at the state, national, or international level?	What external constituents are important to the success of climate-related strategies? How should they be engaged?
Step 1	**Step 2**	**Step 3**	**Step 4**	**Step 5**	**Step 6**	**Step 7**	**Step 8**

Feedback and monitoring to refine business case, strategy elements, and tactics

For example, growing political interest has led DuPont to engage in more climate-related policy discussions with state and federal officials. Changing consumer preferences have prompted Whirlpool (which recently acquired Maytag) to focus more strongly on energy efficiency. According to Casey Tubman, Brand Manager of Fabric Care Products at Whirlpool, "In the 1980s, energy efficiency was number ten, eleven or twelve in consumer priorities. In the last four or five years, it has come up to number three behind cost and performance, and we believe these concerns will continue to grow." Mike Bertolucci, President of Interface Research Corporation, concurs: "Customers are now becoming more aware of the importance of the climate change issue." For companies that sell consumer products, timing is all the more critical given the resources and lead time required to develop effective strategies. According to Jeff Williams, Manager of Corporate Environmental Initiatives at Entergy, "A significant amount of lead time was needed to select, fund, and complete quality projects before realizing CO_2 benefits."

Establish appropriate levels of commitment: Closely related to the issue of timing is the question of how aggressively a company should pursue climate change strategies. Some companies warn that it is important not to get too far ahead of the rest of the business community. In the words of David Bresch, Head of the Atmospheric Perils Group at Swiss Re: "You should always remain one step ahead of the competition. But if you are two steps ahead, you lose the crowd. The ideal is for you to be the leader of the pack and everyone pulling in the same direction." The policy and market impacts of climate change are still very uncertain, and it would be unwise for a single company to pursue initiatives that may not yield financial or strategic benefits. Many companies are concerned about implementing measures in a poorly defined political and market environment. For example, Cinergy (recently merged with Duke Energy) is making some GHG reductions, but CEO Jim Rogers does not believe the company can take aggressive action until there are clear regulatory and market signals. Absent such signals, companies may find it challenging to support GHG reductions and may need to justify action on the basis of short-term initiatives that produce immediate benefits, long-term projections of how climate change may affect their assets and market positioning, and socially and ethically based arguments for "doing the right thing."

"You should always remain one step ahead of the competition. But if you are two steps ahead, you lose the crowd. The ideal is for you to be the leader of the pack and everyone pulling in the same direction."

Influence policy development: Any policy that regulates GHG emissions will set the "rules of the game." Companies are aware that regulation can change the competitive landscape[39] and that future climate policies are likely to favor certain actions, companies, and industries. Relevant factors include the types of emissions regulated, tracking and measurement methods, target setting, and many others. To maintain a measure of control over their future business environment, companies are actively seeking ways to influence policy, and they recognize that credible action can give them greater leverage in that process. As David Hone, Group Climate Change Advisor at Shell, cautions, "To validly have a seat at the table, you have to bring experience. You cannot just take a seat because you are interested."

Create business opportunity: Companies with a history of working on climate change are now trying to shift their focus from risk management and bottom-line protection to business opportunity and top-line enhancements. Goldman Sachs has identified three ways that climate strategies can add value: protecting reputation, enhancing competitive position, and developing new products.[40] Later sections of this book detail some of the specific climate-related business opportunities that companies have identified and are seeking to leverage through early action.

"To validly have a seat at the table, you have to bring experience. You cannot just take a seat because you are interested."

The strategic integration of climate goals with other objectives should be of particular interest to the financial community as analysts seek to develop tools and benchmarks for identifying competitive advantage based on best practices. While this book can help determine whether a company has taken appropriate steps to prepare for carbon controls and new business opportunities, the impact of climate change on company financial valuations needs further study.

Ultimately, sustainable climate-related strategies cannot be an add-on to existing business models, independent of the company's overall competitive strategy. As Linda Fisher, Vice President and Chief Sustainability Officer at DuPont, explains, "We need to understand, measure, and assess market opportunities. How do you know and communicate which products will be successful in a GHG constrained world? How should we target our research? Can we find creative ways to use renewables? Can we change societal behavior through products and technologies? The company that answers these questions successfully will be the winner."

Stage I: Develop a Climate Strategy

This stage involves determining how climate change creates risks and opportunities for a company and outlines steps for developing a strategy to address them.

Step 1: Conduct an Emissions Profile Assessment

The first step in developing a climate strategy is to analyze a company's GHG emissions profile throughout the value chain. This is a fundamental starting point for identifying and prioritizing emissions reduction options, the means to reduce emissions, products and services that may be affected by carbon constraints, and potential strategies that are complementary to the core business. To identify sources, types, and magnitude of emissions, as well as the vulnerability of business lines, companies need a basic awareness of the tools and protocols available to gather such information.

A. Lessons Learned

- Nearly all companies measure direct emissions and most measure indirect emissions. Yet, there is great variability in what emissions are considered. Companies should be aware of the range of possible emissions categories and the extent to which their business activities contribute to each one.
- Companies are evenly split in their use of absolute or indexed measures for tracking and reporting emissions. Absolute measures are necessary for assessing a company's full exposure to carbon constraints, but indexed measures may be useful for setting targets among various divisions or for benchmarking against other companies.
- Companies can measure actual emissions or develop estimates using fuel- or material-based calculations. The former approach may be more expensive and labor-intensive but the latter is complicated by the variety of methodologies that exist for calculating emissions.
- Companies have developed or are working to develop new systems for measuring and tracking emission reductions. These systems can be labor intensive (requiring, for example, energy reporting and verification of third party invoices) and may require further work to be integrated with other information systems (such as SAP[41] and Environmental Management System (EMS) under ISO14000).

B. Emission Types

Nearly all (97 percent) of the companies surveyed for this book have inventoried their emissions of six GHGs: carbon dioxide (CO_2), methane (CH_4), nitrous oxide (N_2O), hydrofluorocarbons (HFCs), perfluorocarbons

(PFCs), and sulfur hexafluoride (SF_6). Many companies have converted all emissions to a carbon-dioxide equivalent (CO_2e) measure using 100-year global warming potential values established by the Intergovernmental Panel on Climate Change (IPCC).[42] Direct and indirect[43] emissions may be included, although there are a variety of ways in which these categories are currently defined.

Direct emissions come from sources owned by the reporting company and generally include emissions from on-site production processes, from the direct combustion of fossil fuels in boilers and furnaces, and from on-site power generation. One area of ambiguity involves emissions from joint ventures and from partial or wholly-owned subsidiaries. Cinergy, for example, only measures direct emissions from facilities in which it has an ownership position and operational responsibility. If a facility meets these criteria, the company assumes responsibility for all GHG emissions and does not prorate on percentage of ownership. Another question revolves around emissions from divested operations. In 2004, DuPont divested itself of its nylon business, Invista®, which generated significant N_2O emissions. The company's decision to subtract these emissions from past performance and baseline measures diminished its overall footprint reduction from 72 percent to 60 percent.

Developing an Emissions Inventory

An emissions inventory is an essential early step in developing a corporate GHG strategy. The World Resources Institute/World Business Council on Sustainable Development (WRI/WBCSD) Greenhouse Gas Protocol Corporate Accounting and Reporting Standard provides a step-by-step guide for quantifying GHG emissions and is used as the starting point for most reporting efforts around the world. Companies can do a Scope 1, Scope 2, or Scope 3 inventory. Scope 1 includes direct emissions; Scope 2 includes indirect emissions from the consumption of purchased electricity, heat, or steam; and Scope 3 includes other indirect emissions from upstream and downstream sources, as well as emissions associated with outsourced or contract manufacturing, leases, or franchises not included in Scope 1 or Scope 2. The WRI/WBCSD Protocol includes guidance for identifying relevant source categories and calculation tools for emissions from particular source categories.[44]

The 77 percent of survey respondents that measure indirect emissions—that is, emissions that do not directly occur at the reporting company's facility—use a variety of approaches. The most commonly reported sources of indirect emissions are electricity, heating or cooling, and steam purchased from a third-party provider. Cinergy, on the other hand, does not track power purchases from third party vendors, as such variation raises questions of double counting if the seller is also counting these as direct emissions.

Other companies measure emissions generated by the use of their products (defined as Scope 3 emissions by the WRI/WBCSD reporting and accounting protocol; see "Developing an Emissions Inventory" on this page). For Whirlpool, the use of its home-appliance products constitutes 93 percent of the company's GHG profile and is the primary focus of reduction efforts. Alcoa, on the other hand, does not consider its product use in its emissions profile.

A small number of companies (such as IBM, Interface, and several financial-services firms) account for emissions from material transport, business travel, and/or commuting. Swiss Re, for example, generates 43 percent of its emissions profile from business travel (direct emissions and indirect office electricity use account for the remaining 13 and 44 percent, respectively). Cinergy, on the other hand, does not count upstream emissions from the mining and transport of coal.

One final issue concerns emission credits from biological carbon sequestration, something that Cinergy includes in its inventory. The company identifies test plots and measures tree volumes, underbrush, and soils for carbon content. The measurements are repeated at regular intervals and resulting data are extrapolated to the entire acreage of plantings. The company states that this process yields a 95 percent statistical level of confidence.

C. Emission Metrics

GHG emissions can be reported in a variety of forms. To gain a clear sense of a company's exposure to possible climate-change policy, absolute measures are necessary. But to track performance relative to other economic or production goals, or to competitor benchmarks, emissions can be normalized with another measure such as dollars of revenue or units of production.

"There are regional differences in approaching the issue that require a company to have both a global and regional focus."

Survey respondents are evenly split between absolute or indexed measures for tracking and reporting progress on GHG emission reductions (48 and 52 percent respectively).[45] The most common metrics include: total tons CO_2e; tons CO_2e per unit of product; energy consumption per unit of product; and total energy consumption.[46] (See Table 3)

Table 3

Most Common Metrics for Measuring GHG Emissions*

Metric	Percent
1. Total tons of CO_2e	73 percent
2. Tons of CO_2e per unit of product	50 percent
3. BTU (energy consumed) per unit of product	39 percent
4. Total BTUs	35 percent

*Many companies use more than one metric.

A minority (12 percent) of survey respondents use a combination of absolute and indexed metrics, tailoring emissions measurements as appropriate for particular goals or reporting units and regions. Daniel Gagnier, Senior Vice-President of Corporate and External Affairs at Alcan, explains that despite the general trend to evaluate reductions from a global perspective, "there are regional differences in approaching the issue that require a company to have both a global and regional focus." Shell decided that setting a universal standard would be impractical because of its size and multinational focus. Even though the company reports an absolute target externally, it gives individual business units the flexibility to use indexed measures for internal reporting. This approach is particularly popular with units that have significant growth opportunities.

D. Emission Measurement Tools and Techniques

Some companies measure actual emissions, while others estimate emissions using fuel-based calculations (based on methodologies such as those created by WRI/WBCSD, the European Union Emission Trading Scheme (E.U. ETS), the U.S. Department of Energy (DOE) and others). The difference depends, in part, on the complexity of the task. Companies with many emission sources or extremely hostile stack environments prefer to avoid on-site measurement due to the cost of purchasing, installing, maintaining, and replacing monitors. Cinergy measures CO_2 directly at generating units equipped with continuous emissions monitors (CEMs).[47] For units not equipped with CEMs, estimates are calculated using the energy (BTU) value of the fuel consumed multiplied by its carbon intensity (pounds of CO_2 emitted per million BTU) as provided through the DOE's Energy Information Administration (EIA) 1605(b) reporting program.

The majority (62 percent) of survey respondents have developed new information systems or monitoring equipment to track GHG emissions. The functionality of these measurement systems varies considerably by company: some use highly sophisticated, web-based database tools, while others are still in the development process.

All companies recognize the importance of emissions measurement and tracking. Alcoa, for example, considers the development of its internal web-based GHG information system a major step toward achieving its climate goals. Its centralized system currently includes detailed process and energy consumption information for 41 facilities worldwide, including four power generation facilities, nine alumina refineries, and 26 smelters. Alcoa's system uses the methodology of the E.U. ETS to calculate emissions and sweeps databases every evening to download process and production data. Designated individuals at each plant are responsible for manually entering energy consumption data on a monthly basis and reminders are issued automatically to ensure that data for all facilities are available as soon as possible after the end of each month.

Some companies have been able to incorporate GHG tracking into integrated performance measurement systems like SAP, allowing them to link emissions reductions to financial measures. Ruksana Mirza, Vice President of Environmental Affairs at Holcim (U.S.) Inc., states that the company's SAP enterprise resource planning platform is linked to a CO_2 module that automatically calculates monthly emissions using relevant operating information (e.g., production volume, energy consumption, fuel type, etc.).

At DuPont, progress on GHG reductions is tracked at the business-unit level through the Corporate Environmental Plan, a database that captures annual performance information on waste, GHG and other emissions, and energy use at company facilities worldwide. Energy-related emissions are calculated based on fuel consumption according to the WRI/WBCSD GHG reporting protocol. The current system requires data inputs from direct metering of gas consumption, invoices for other fuel purchases, reconciliation to inventories, and emissions factors for a variety of fuels. Process emissions are reported separately and indirect emissions are calculated based on localized information. All of this information is collected once per year in the corporate database.

As these examples illustrate, tracking GHG emissions can be complicated and, at times, labor intensive. While some companies are generally satisfied with the performance of their respective systems, many see room for improvement. For example, despite having tracked emissions since 1991, John DeRuyter, Principal Consultant, Energy Engineering at DuPont, still believes the "biggest headache is in capturing and reporting data, particularly energy reporting and verification of third party invoices." There is no link with the company's SAP system, which would be desirable but is currently prohibitively expensive.

Other companies point to a need for better measurement tools in the future. According to Steve Willis, Director of Global Environment, Health and Safety at Whirlpool, a data management system and international GHG conversion factors are the most significant requirements for implementing a climate strategy.

Step 2: Gauge Risks and Opportunities

Emissions alone do not reveal a company's exposure to carbon constraints. Companies must also consider potential impacts on product and service lines.

The next step in climate-strategy development is consideration for how operations and sales may be affected—both for the positive and the negative—by climate change-related factors and, as a result, how such factors may alter competitive positioning. As part of this analysis, companies should consider their emissions profile relative to industry peers, the industry's position relative to other sectors, potentially relevant future regulatory developments, trends in input costs, and potential changes in customer preferences. Identifying risks and opportunities must flow from an understanding of the company's current and future GHG footprint in the context of a current and future carbon-constrained society and economy.

A. Lessons Learned

- Benchmarking is geared towards gaining information on best practices, as well as gaining the strategic benefits that come from being identified as a leader on climate change.
- In assessing product and service line vulnerabilities, companies begin with a focus on risk management and bottom-line protection.
- With time and experience, companies then shift their strategies for addressing climate change to emphasize business opportunities and top-line enhancements.

B. Benchmarking

Seventy-three percent of survey respondents report that they benchmark against other companies on their climate-related performance. In general, these companies report that they benchmark against the other companies in their own industries, but many also report that they identify singularly-visible companies that have gained exposure for their climate activities.

The goals of benchmarking activities are to identify best practices for addressing climate change, as well as managing reputation and industry status on the issue. Benchmarking can help protect the company against being identified as a laggard, but more importantly, can help the company gain the benefits of standing out as a leader. The experience of the Pew Center is that an increasing number of companies turn to benchmarking of industry peers in target setting, especially in certain industries and for second-round goals (see discussion of benchmarking in Step 4 on page 25). In the former case, a company that is labeled as a low performer may be susceptible to costly criticism in the press or from NGOs. In the latter case, however, securing a first-mover advantage in addressing climate change can create opportunities and garner recognition (such as through rankings by *Business Week*, *The Financial Times*, Ceres, and others). Securing this recognition requires efforts at external outreach, which are discussed in Step 8. Finally, benchmarking can inform the need for collective industry action, especially if an industry wishes to achieve sufficient reductions on a voluntary basis to reduce pressure for onerous regulation.

C. Risks from Operations, Products, and Service Lines

In terms of assessing product and process line vulnerabilities to carbon constraints, companies generally begin with a focus on risk management and bottom-line protection. Cinergy, for example, is concerned about the impact of climate-change regulation on the value of its existing and future energy producing assets. When the company first completed its GHG profile for the year 2000, it was clear that the majority of emissions came from legacy electric-generating units. Because new generating capacity has an expected lifespan of 40 to 50 years or more, Cinergy is particularly sensitive to uncertainty surrounding future climate policy as it relates to strategic investments. Therefore, the company is working to develop new technologies for reducing emissions from its coal-fired assets and sees increased value in the nuclear capacity acquired through its recent merger with Duke Energy.

Ultimately, the most effective climate-related strategies connect GHG reductions with a company's core business strategy.

For Alcoa, one core business concern centers on aluminum production costs. As an energy-intensive basic materials company, securing reliable, low-cost, long-term energy sources is among its most pressing strategic priorities. Climate policies threaten to alter the economics of critical energy inputs. Thus, against the backdrop of global climate policy trends, the emissions profile of existing and new energy sources has been a focus of its strategy.

D. Product and Service-Line Opportunities

While risk management can be a starting point for addressing climate-related vulnerabilities, with time and experience companies shift their climate-related strategies to emphasize business opportunities and top-line enhancements. In fact, the mere presence of risk from GHG intensive operations, products, and services signals the potential for business opportunities based on GHG efficiency. Companies need to assess whether and how demand for their current and future product and service lines may be enhanced by climate-related developments.

Ultimately, the most effective climate-related strategies connect GHG reductions with a company's core business strategy. This can be done in a variety of ways. One way is through operational improvements. For example, instead of flaring methane gas in its exploration and refining operations, Shell now captures the gas and either pumps it back underground to enhance well production or feeds it to nearby facilities for power production. When the economics are right, the methane can be converted into liquefied natural gas (LNG), a major potential growth area.

Another approach is to find new uses for existing product lines and to develop new products to satisfy emerging market needs. For example, DuPont developed a special grade of Tyvek® house wrap for European customers; this wrap reduced energy use and CO_2 emissions and lowered heating bills. DuPont engineers also work directly with the company's business customers to help them reduce energy consumption. This strategy not only delivers higher value, it enhances DuPont's relationships with its customers and may be rewarded by larger or longer-term contracts.

Swiss Re is also looking for ways in which to augment its existing activities to create opportunities from climate change. For example, the company channels investments in its sustainability portfolio into a number of sectors, including alternative energy, water, and waste management/recycling. More specifically, the company seeks opportunities representing medium to high risk-return profiles in: infrastructure investments such as wind farm, biomass, and solar projects; investments in publicly quoted, small- to medium-capitalized growth companies, and; cleantech venture capital investments, representing the highest risk-return profile. As tightening policy frameworks increase demand for such projects, the company's investment strategy is beginning to pay off. The portfolio's market value rose substantially in 2005 thanks to strong share performance, as well as new investments.

Yet another way to create synergies between climate and business strategy is through acquiring assets that balance a company's portfolio. For example, Cinergy's emissions profiling and assessment of likely future regulatory scenarios pointed to increased value in nuclear capacity, which it gained through its recent merger with Duke Energy. Any form of GHG regulation will favor electricity from no-carbon and low-carbon sources over time, signaling potential advantages for operators of nuclear plants. Moreover, this potential opportunity may grow for utilities like Duke, Exelon, Entergy, and others that have particular expertise in permitting, building, and operating nuclear capacity.

Alcoa found that future climate policies may create market opportunities by expanding aluminum recycling. Considering that aluminum produced from recycled materials requires only five percent of the energy needed to make primary aluminum and that energy prices will likely continue to rise, the company has pledged that 50 percent of its products, other than raw ingot sold to others, would come from recycled aluminum by 2020. Increasing recycling rates is among the more significant long-term strategic opportunities for the company. Another is the expected boost in demand for aluminum as a material in lighter-weight vehicles, and the company is continuing to make progress into this area. For example, Alcoa developed "Dura Bright" commercial truck wheels that are lower mass than conventional wheels and, as an added marketing advantage, don't require polish or scrubbing. According to the company, a 10 percent reduction in vehicle weight typically yields a 7 percent reduction in GHG emissions.

Similarly, Whirlpool's business opportunity lies with consumer choices. As most lifecycle GHG emissions from home appliances come from their use rather than production, the company's primary focus is on appliance efficiency. Whirlpool expects mounting awareness of climate issues and rising energy costs to drive consumer demand toward less energy-intensive products and therefore is leveraging its core competencies to continue bringing the most energy-efficient appliances to market.

But, going even further, some companies have focused their energy and efforts into fundamental technology shifts. DuPont, for example, has identified the most promising growth markets in the use of biomass feedstocks that can be used to create new bio-based materials such as polymers, fuels and chemicals, new applied Biosurfaces, and new Biomedical materials. One promising development is Sorona® polymer. In a joint venture between DuPont and Tate & Lyle PLC set to go on-line in the third quarter of 2006, the company will begin producing 1,3-propanediol, the key building block for the new polymer, using a proprietary fermentation and purification process based on corn sugar. This bio-based method uses less energy, reduces emissions, and employs renewable resources instead of traditional petrochemical processes. Another promising development for DuPont is the 2006 creation of a partnership with BP to develop, produce, and market a next generation of biofuels. The two companies have been working together since 2003 to develop products that will overcome the limitations of existing biofuels. The first product to market will be biobutanol, which is targeted for introduction in 2007 in the U.K. as a gasoline bio-component. This biofuel offers better fuel economy than gasoline-ethanol blends and has a higher tolerance to water contamination than ethanol.[48] Both of these developments represent the new direction in which the company is headed—one that significantly reduces the company's environmental footprint. According to Uma Chowdhry, VP of Central Research and Development, this is not a subtle shift, but rather a significant change in product lines and research focus for DuPont. She is hoping that DuPont will soon be known for leading the industrial biotechnology revolution and predicts that over 60 percent of DuPont's business will stem from the use of biology to reduce fossil fuel use in the next few decades.

New Products Reduce Carbon Dioxide Emissions

Jeff Hawk, Director, 787 Government and Certification, The Boeing Company

In 2008, The Boeing Company will begin delivery of a new airplane—the 787 Dreamliner—that reduces CO_2 emissions by approximately 20 percent compared to today's similarly sized airplanes. In addition to new, more efficient engines, Boeing also redesigned the airframe to be more efficient by streamlining its aerodynamic shape and significantly increasing use of lightweight composites. Airline interest in the airplane has already generated a record number of firm orders—more than 375 since the plane was launched about two years ago. Airplane design decisions are based on market requirements including passenger loads and route structure; research to reduce CO_2 emissions from the 787 began in the mid-1990's. After studying the market with customers, Boeing decided to design a mid-sized airplane (200 to 300 passengers) that could travel as far as today's bigger jets while reducing CO_2 emissions through lower fuel use. By designing an airplane that can travel efficiently on long-range routes with fewer passengers, the 787 can help to eliminate stops at "hub" airports, creating more direct flights and reducing extra flight miles, saving fuel, CO_2 emissions, and passengers' time. While aviation represents less than 3 percent of worldwide CO_2 emissions (according to the IPCC), it is a growing sector of the economy. Boeing believes it is important to continue environmental improvements, especially CO_2 and noise reductions, to meet future expectations.

Step 3: Evaluate Options for Technological Solutions

After developing an emissions profile, the next task is to evaluate options for reducing emissions. This step is often conducted in an iterative fashion with goal setting. Some companies set goals and then search for ways to achieve them. Others consider their options for reducing emissions and then set goals accordingly. The precise ordering is a matter of individual management style.

A. Lessons Learned

- Many companies were able to identify a variety of low-cost options for reducing their GHG emissions. These "low-hanging fruit" opportunities often include behavioral or technological changes that improve efficiency and reduce energy consumption.
- A few companies developed breakthrough technology solutions that facilitated a dramatic reduction in their GHG footprint. Such "silver bullet" opportunities are often the focus of new technology development but have also been realized in existing operations.
- Over the long term, companies can develop and fine tune programs to implement more challenging solutions. This can involve technical efforts aimed at plant optimization as well as organizational efforts related to information sharing and internal consulting.
- Companies typically distinguish on-system options for reducing emissions from off-system opportunities. On-system reductions involve projects within a company's operations. Off-system reductions can include forest sequestration projects, purchased offsets, sourcing offsite renewable energy, and others. Public and private benefits and costs for these reductions vary.
- Ultimately, the goal is to find ways to reduce GHGs in a manner that supports other business objectives. This can involve linking emission reductions to improved operations, finding new markets for existing products, or creating new products to serve emerging markets.

B. Low-Hanging Fruit

Companies in this book have found myriad low-cost or low-risk, easily identifiable solutions for lowering their emissions profiles. This is particularly true for companies that have not actively pursued reductions in the past. Such "low-hanging fruit" may include simple energy efficiency initiatives, behavioral changes, or process improvements. For example, the first step in Swiss Re's three-tiered approach to reducing GHG emissions involved turning down heating and cooling in company offices and turning off lighting systems during non-working hours. As a second step, the company has focused on small investments, such as motion sensors and compact fluorescent light-bulbs, and on reducing emissions from business travel by curtailing short-distance trips for internal meetings and by providing employees with the latest telephone or video conferencing technology. The final tier of Swiss Re's approach involves refurbishing company-owned property and buildings by, for example, replacing cooling towers, generators, insulation, or windows.

Many electric-generating companies have undertaken efficiency improvements at individual power plants to produce more electricity per unit of fuel input. Boosting output at non-GHG emitting nuclear, hydro, and landfill-gas facilities can be an effective approach to reducing emissions as well. In sum, though the specific opportunities may differ, all companies should identify "low-hanging" options.

C. "Silver Bullets"

Some companies have achieved dramatic GHG reductions by implementing a single initiative that significantly altered their emissions profile. For example, Shell managed a sizable portion of its pre-2002 emissions by reducing the venting of associated gas (methane) from its exploration and production facilities. In many cases, "silver bullet" initiatives require innovation and investment in improved processes. Most of Alcoa's GHG-reduction efforts have involved cutting perflourocarbon (PFC) emissions by reducing the number of anode effects in the aluminum smelting process (see "Anode Effect: An Overview" on page 105). DuPont's "silver bullet" involved reducing emissions of two potent GHGs—N_2O and HFC-23 (an unintended byproduct of producing HCFC-22, a common refrigerant)—through two discrete process technologies.

Clean Energy for a Low-Carbon Future

Bill Gerwing, Director Health, Safety, and Environment, BP America

BP believes that power generation and a clean environment are not opposing goals. That belief is at the heart of BP's plans for a Hydrogen Power Project in Carson, California—in what would be the company's second industrial-scale hydrogen power project—designed to generate electricity and reduce greenhouse gas emissions by capturing carbon dioxide (CO_2) and storing it safely and permanently. The result would be California's cleanest new power plant. The project would combine a number of existing industrial processes to provide a new option for generating electricity without significant CO_2 emissions. Petroleum coke produced at California refineries would first be converted to hydrogen and CO_2 gases with around 90 percent of the CO_2 captured and separated. The hydrogen gas would be used to fuel a power station capable of providing the California power grid with 500 MW of electricity, enough to power about 325,000 homes in Southern California. At the same time, about four million tons of CO_2 per year will be captured, transported and stored in deep underground oil reservoirs where it will enhance existing oil production. When completed, the Carson Hydrogen Power Project would be the largest hydrogen-fired power generation facility in the world and would have the lowest CO_2 emissions in the world for an integrated gasification combined cycle (IGCC) plant.

Even if such big-impact opportunities do not exist at present, companies are working to make them a reality in the future. Alcoa is developing a new smelting technology based on an inert anode which would eliminate consumable carbon anodes and related PFC and CO_2 emissions. For Cinergy, coal IGCC technology combined with carbon capture and sequestration (CCS) holds promise for reducing future emissions. The company has been involved in IGCC since the early 1990's when it built one of the first demonstration plants in the United States; it is now working with GE Energy and Bechtel Corporation to study the feasibility of a commercial scale (600 MW) IGCC generating station with CCS.

D. Ongoing Reductions

To continue making GHG reductions, companies must sustain their efforts over the long term. For example, Alcoa and Shell are conducting ongoing efficiency assessments at operating plants to generate recommendations for operational, equipment, and behavioral changes. They have also facilitated information sharing about promising energy practices between plant locations. Both companies provide technical support and access to further resources as needed. Similarly, DuPont is focusing its ongoing GHG-reduction efforts on more capital-intensive measures that affect yield, capacity, and utilization; process changes; combined heat and power; and modern heat management using insulation, steam traps, waste heat recovery, and modern motors.

Capital investments to reduce energy consumption often meet resistance because they are not viewed as "sexy" or compelling. If the pool of resources is dwindling, the certainty of returns in energy-efficiency projects can actually become a liability.

One challenge to sustaining climate-related strategies is that they must compete with other initiatives for funding. According to John Carberry, Director of Environmental Technology at DuPont, capital investments to reduce energy consumption often meet resistance because they are not viewed as "sexy" or compelling. If the pool of resources is dwindling, the certainty of returns in energy-efficiency projects can actually become a liability. DuPont, for example, has ruled out lowered hurdle rates, internal carbon shadow pricing, or setting a budget for energy efficiency projects: "The problem is that when we pitch 20 percent return with 99 percent certainty on energy, we lose to a marketing group pitch of 40 percent return with 60 percent certainty," says Carberry. The choice to create internal price supports for emissions-reduction efforts is a strategic decision for companies; absent such supports, climate-related projects must show positive returns relative to other initiatives if they are to offer value to the company.

E. On-System versus Off-System Reductions

Not all GHG reductions are made at the facility level or even within the company: Cinergy, for example, intends to achieve up to one-third of its emissions reductions off-system. Off-system reductions can include forestry projects, end-user efficiency programs, and research and development projects. By contrast, Cinergy's on-system projects target direct emissions from smoke stacks and vehicle tailpipes, or methane emissions from the company's natural gas distribution system.

Swiss Re plans to achieve 15 percent of its reduction target through actual facility reductions and the remaining 85 percent through off-system investments in the World Bank Community Development Carbon

Fund. However, sourcing emissions credits through external means is not always easy. DuPont has found that cost-competitive alternative energy projects are relatively scarce and difficult to identify. The company has been able to source about 5 percent of its energy from renewable sources but hopes to increase that amount. These two examples highlight a distinction between purchasing emissions offsets that are not directly related to a firm's activities (the Swiss Re/World Bank example) and sourcing renewable energy for a firm's own use (the DuPont example). Each has very different public and private benefits and costs.

Step 4: Set Goals and Targets

The companies in this book have made a wide range of commitments to reduce GHG emissions, the specifics of which differ in such aspects as timetable, objectives, baseline year, and types of emissions covered. For example, DuPont's goal of reducing GHG emissions 40 percent below 1990 levels by the year 2000 was set in 1994. That target was met in 1999, and the company established a new goal to reduce net GHG emissions 65 percent below 1990 levels by 2010. Whirlpool's target, set in 2003, calls for reducing total GHG emissions from global manufacturing, product use, and disposal by 3 percent from a 1998 baseline by 2008, while also increasing sales by 40 percent over the same period.

Goals and targets need not be limited to GHG reductions but can include strategic initiatives and adaptation strategies. As noted earlier, goals can be based on identified emissions-reduction opportunities or set as stretch goals. Most companies establish short- and long-term goals in an iterative fashion and in a way that is aligned with their strategic objectives. Ultimately, these goals must fit the company's capabilities, culture, and business model.

A. Lessons Learned

- Companies adopt targets for a variety of internal and external reasons. Some have identified climate change as a significant strategic issue for the future while others have been prodded by shareholders or other external constituencies.
- Companies cite three primary sources of motivation: cost savings, social responsibility, and reputation enhancement. The latter is linked to an expectation of enhanced ability to foresee and influence future regulation.
- Most companies develop goals by analyzing risks and opportunities in their many business units. Those that have achieved the most dramatic GHG reductions set stretch goals beyond what their original analysis indicated was possible. Many of these companies subsequently achieved their goals before the target date and set new, more ambitious ones.

- Many companies establish both energy-efficiency and GHG-reduction targets. While efficiency improvements often yield near-term financial benefits, the value of GHG reductions is more difficult to quantify and serves longer-term objectives. As a result, efficiency programs are more likely to be considered strategic and proprietary, while GHG reductions may be difficult to connect to a company's bottom line.
- In making the business case for climate-related strategies, companies typically focus on the quantifiable financial benefits of energy efficiency projects, the less quantifiable reputational and organizational benefits of "doing the right thing," and scenario planning that highlights the future likelihood of, and impact from, carbon constraints.
- Companies are also developing other goals. Examples include sourcing renewable energy, reducing solid waste, eliminating all waste, increasing use of hybrid biofuels and vehicles, and others.
- Finally, companies are developing adaptation strategies to be prepared for the physical risks associated with climate change.

B. Motivating Factors

A company's motivations for taking action can be influenced by corporate history and culture, core competencies, or the competitive environment. Many companies in this study first became involved through a narrowly focused internal initiative. Cinergy's efforts began in the early 1990's after the company commissioned a study on the feasibility of adopting an internal CO_2 cap. Shell had been watching the climate change issue since the early 1990's through its Issues Management Team within Corporate Affairs. In 1998, Jeroen van der Veer (then a group managing director and now Shell CEO) championed a more formal study of climate change and its potential impact on the company's businesses globally. DuPont's actions were foreshadowed by its experience with stratospheric ozone depletion in the 1970s and 1980s. When the IPCC issued its first assessment report in 1990, DuPont's (former) CEO Ed Woolard saw a familiar scenario playing out and directed the company to become an early adopter of GHG reductions.

Interest in climate-related strategies can also be motivated by outside parties. While Cinergy's CEO and management team were interested in climate action from the early 1990's, concurrent shareholder resolutions in 2002 and 2003 helped the company take the final step. Dialogue with shareholders resulted in a plan to disclose risks related to climate regulation, and Cinergy formally announced its internal GHG-reduction program in September 2003.

While the specific impetus for each company varies, three overarching drivers emerged from the survey: cost savings, social responsibility, and reputation. These drivers are linked by a common desire to ensure the long-term success of the organization and are discussed in more detail below. It should be noted, though, that as a company fulfills its goals in these areas and gains knowledge of the issue, the motivations then shift toward leveraging climate-related market changes for competitive advantage. Companies just starting the strategy process for the first time should tap this motivation from the start.

Figure 2

Motivations for Undertaking Climate Action

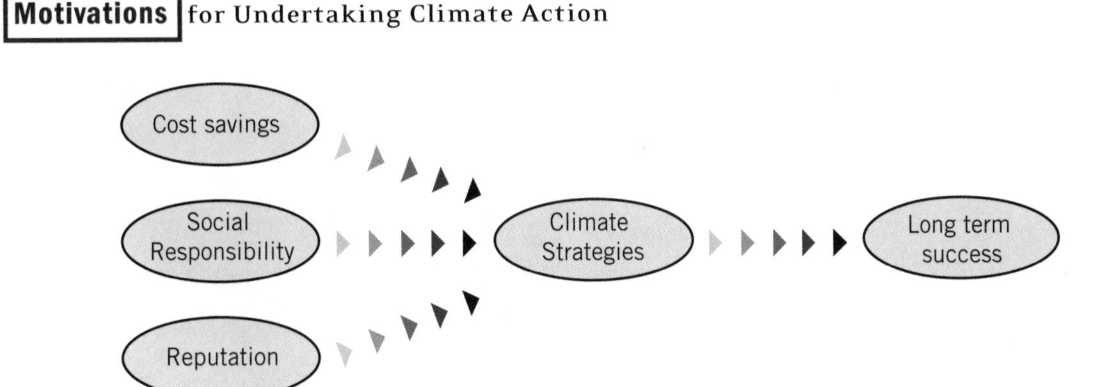

Cost savings: As noted previously, near-term cost savings are generally realized through improvements in energy and operational efficiency (see Figure 3). Survey respondents rank efficiency improvements as the most prominent measure of success (see Figure 4) and the action that most often provides bottom-line benefits (see Figure 5).

Social responsibility. Although social responsibility (often characterized as the desire to "do the right thing") ranks low in terms of generating short-term bottom-line benefits (see Figure 4), companies consider it a primary motivator and see early action on climate as consistent with their corporate values (see Figure 3). For example, DuPont cites its culture of science, safety, and environmental responsibility, while Cinergy points to its cultural values and a history of responsibility, transparency, and stakeholder engagement. For Alcoa, climate strategy is part of the company's sustainability efforts, which in turn feed into overall corporate goals. Whirlpool draws a close connection to its Midwestern roots, which foster a strong belief in corporate citizenship. At Whirlpool, according to Mark Dahmer, Director of Laundry Technology, one of the company's core principles is that there is "no right way to do a wrong thing."

Reputation. Companies are also motivated by the desire to protect or enhance their reputation. Remaining inactive on climate change can expose the firm to criticism and negative press, particularly when industry peers are taking action. Conversely, meaningful action can create good will with investors, customers, regulators, and communities. For Swiss Re it is critical to have a voice in social and political debates over climate change. So, while recognizing that its emission reductions amount to merely a "rounding error" compared to other companies, Chris Walker, Managing Director and Head of Sustainability Business Development says, "We need to do this if we are going to be seen as credible."

Figure 3

Drivers of Climate-Related Strategies

How important were the following external drivers in leading your company to pursue its climate-related strategy?

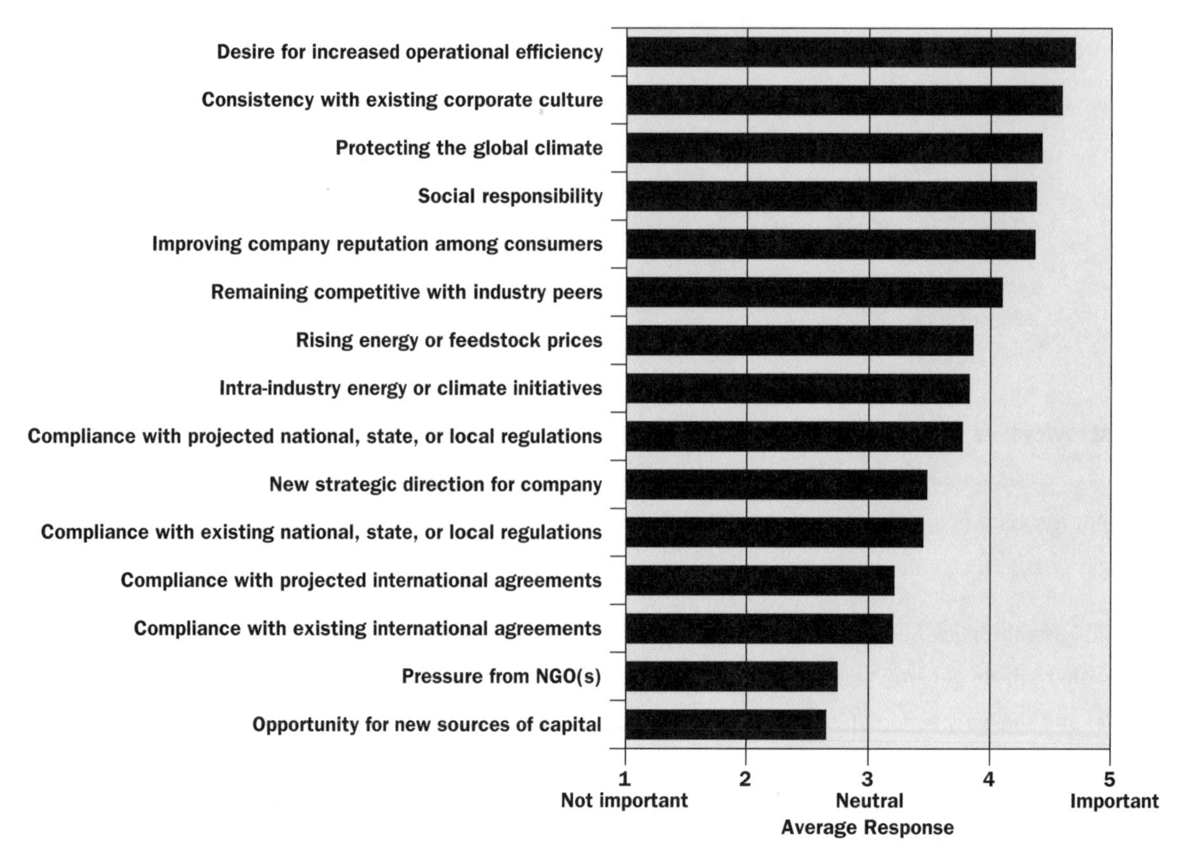

Total Respondents: 30

Alcoa recognizes that a good environmental reputation enhances its ability to site and build new plants. Whirlpool wants its customers to see the company as a source of energy-efficient and environmentally friendly products. Shell has long been motivated by the belief that a leadership position would allow the company to foresee and possibly influence government policy. Similarly, other survey respondents consider the ability to anticipate future regulations to be a critical measure of success (see Figure 4).

C. Developing Climate Goals and Targets

Understanding the context in which a company first takes note of climate change can help inform the development of meaningful goals. Companies in this book tend to be introspective, looking inward at capabilities and interests when establishing targets. Whirlpool, for example, began by soliciting input from each of its product

Figure 4

Once begun, how important are the following measures of success in undertaking your climate-related strategy?

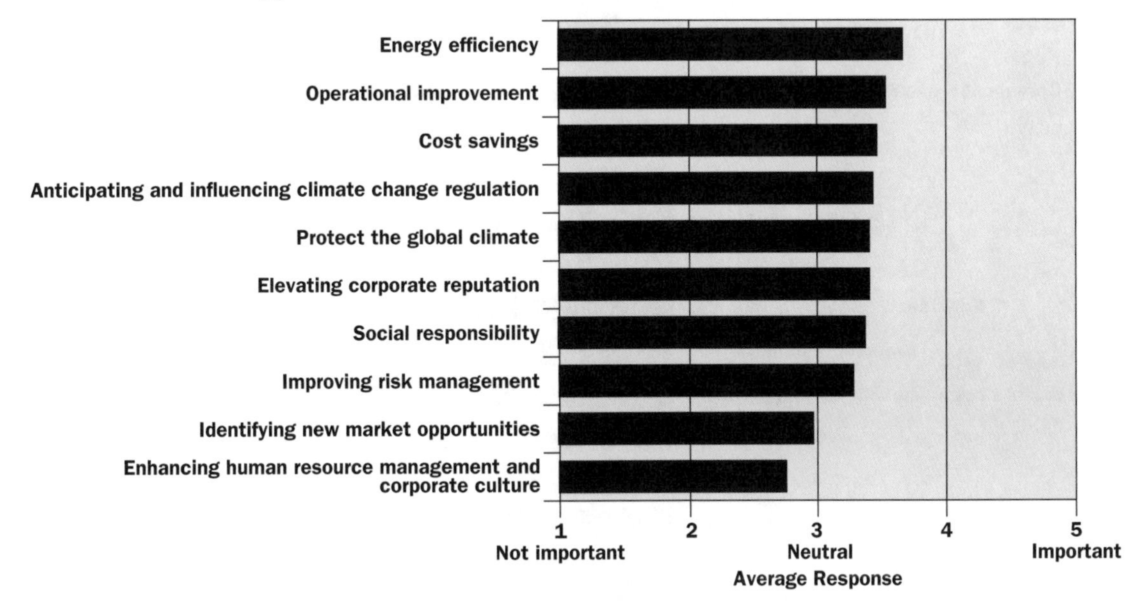

Total Respondents: 30

groups and compiling data on projected sales volumes, consumer use, product turnover, and plans for introducing new, more efficient models. Total energy consumption was then calculated over the average life of each product and converted to GHG emissions using country-specific factors. (See Step 2 on page 14 to learn more about the use of benchmarking for target setting).

Several companies solicited opinions from individual business units but then pushed further, creating a stretch goal to make significant progress. Craig Heinrich, leader of the global energy team for DuPont's Titanium Technologies division explains, "You need the tension of a very challenging goal. Inspirational goals call an organization to act beyond conventional boundaries...an easy goal fails to challenge the creative potential of the organization."

Several companies have met or exceeded internal targets before the stated deadline. Alcoa reached its 2010 emission-reduction goals seven years early, and Entergy not only met its goal of stabilizing CO_2 emissions at 2000 levels, but reduced emissions by an additional 21 percent as of the end of 2004.[49] Andreas Schlaepfer, Head of Internal Environmental Management at Swiss Re, believes that for non-manufacturing companies like Swiss Re, substantial reductions from building-related conservation efforts are quite easy: "If you've never focused on energy efficiency before, achieving a 30 percent reduction is simple."

Figure 5

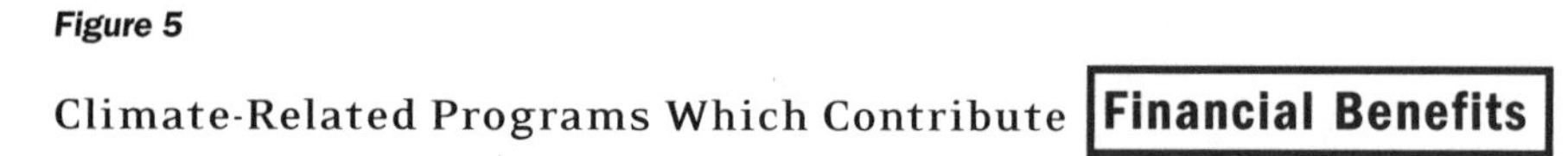

Climate-Related Programs Which Contribute **Financial Benefits**

*Please indicate which are providing positive returns to the bottom line.**

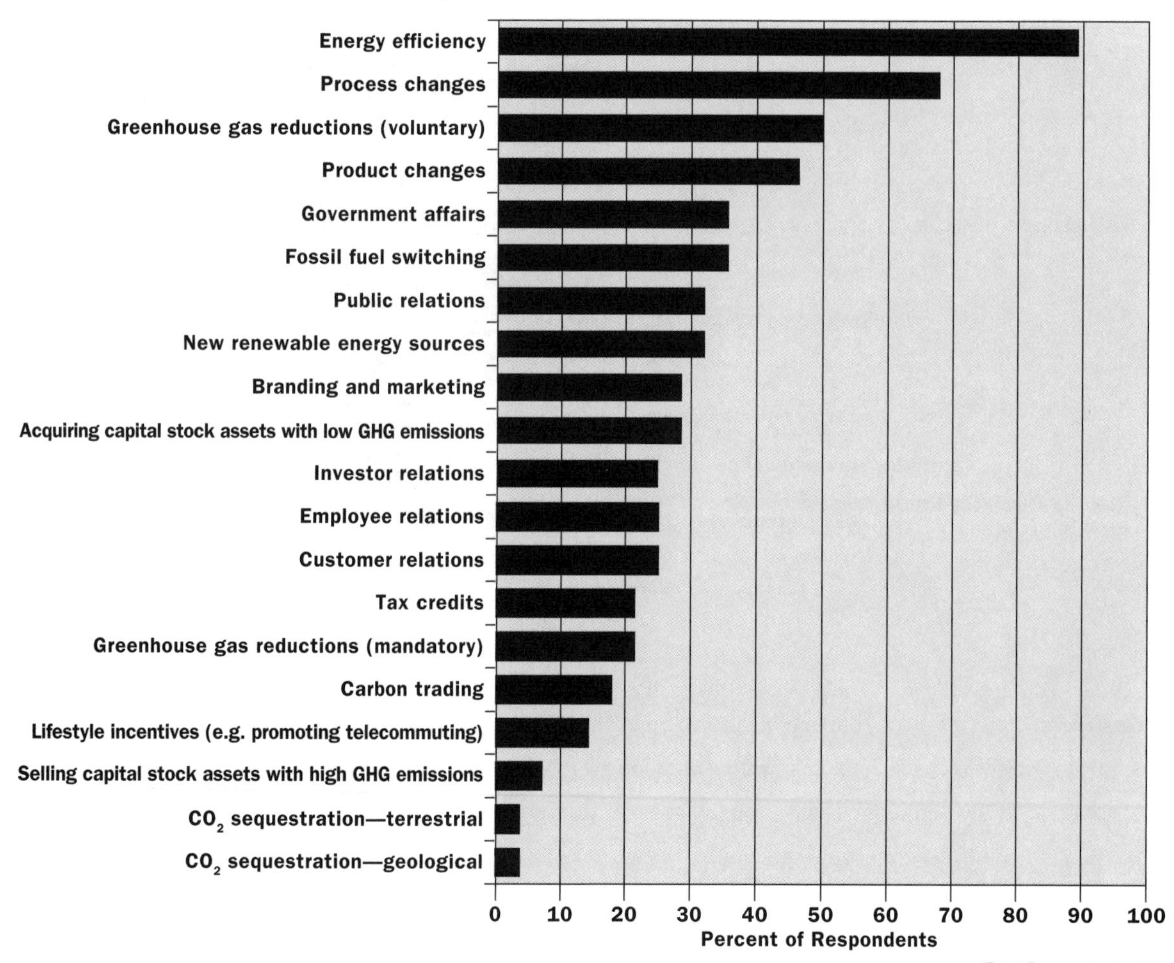

Total Respondents: 28

*Not all of these program elements are relevant to all survey companies, so the responses may be skewed. For example, geologic sequestration only applies to oil and gas companies, chemicals, and a few others at this stage of development of the technology.

Interviewees warn, however, that companies should avoid creating one universal reduction target. They suggest it is best to develop a set of diverse targets across different business units that all contribute toward the overall corporate goal. Interface's Bertolucci advises against "requiring standardized implementation programs in a diverse, decentralized culture." DuPont, for example, expects the output of its Titanium Technologies division to double from 1990 levels by 2010. Because energy comprises a significant percentage of the selling price of titanium dioxide (TiO_2), this creates a significant challenge for meeting the company's energy-efficiency and climate-change goals. The division has been tasked with the goal of increasing energy use by only 40 percent. As a result, other divisions will be expected to make deeper cuts. Alcoa also allows diversity in its divisional targets, but this variation primarily reflects geographic differences. Local management is permitted to determine the company's official position on climate policy within each country based on local circumstances.

D. Differentiating GHG-Reduction and Energy-Efficiency Targets

While energy-efficiency and GHG-reduction targets eventually need to merge into an integrated approach, companies tend to treat them as separate goals in the short run. Among survey respondents, 72 percent have established energy-efficiency goals, whereas 77 percent have established GHG-reduction goals. Of the former group, 100 percent have reached their energy-efficiency targets and 66 percent have established new, more ambitious ones. In contrast, only 60 percent of companies that have adopted GHG-reduction goals have met them. A close look at three aspects of these goals helps to explain the difference.

"You need the tension of a very challenging goal. Inspirational goals call an organization to act beyond conventional boundaries... an easy goal fails to challenge the creative potential of the organization."

First, energy-efficiency goals have a longer **history**. Energy efficiency was first discussed by some surveyed companies as early as 1970, and the average date for setting efficiency targets was 1998. One company states that energy efficiency "was always an objective." By comparison, GHG-reduction goals entered the discussion much later. At the earliest, survey respondents did not discuss them until 1990, and the average date for setting these types of targets was 2000. In several cases, these initiatives have not yet run their course because many companies set initial deadlines of 2006 or beyond.

Second, energy-efficiency and GHG-reduction goals tend to have a different **focus**. Energy-efficiency strategies are generally directed at discrete, energy-intensive processes, requiring units with operational responsibility to make local decisions regarding improvements. Moreover, they often offer a return on investment. GHG-reduction goals, on the other hand, are usually articulated at the corporate level as an "X percent reduction by date Y" without well-defined ways to filter this goal down to individual business units.

Third, energy-efficiency and GHG-reduction goals differ in their direct, near-term impacts on a firm's **bottom line**. For many companies, energy efficiency is seen as an important strategic business issue. GHG reductions, on the other hand, are typically viewed as an initiative of the EHS department, sometimes carrying an (actual or perceived) upfront cost rather than, in the absence of regulation, providing a competitive advantage. All companies publicly report progress toward meeting GHG-reduction goals. But 17 percent of companies that have energy-efficiency targets do not publicize information about their performance results because this information is considered proprietary.

Quantifying bottom-line risks and rewards is important in generating internal support for GHG-reduction strategies. As David Steiner, Vice President of Government Affairs at Maytag, plainly states, the "company must

Making the Business Case for Climate Action[50]

With a relatively straightforward body of measurements and analyses, almost any company can develop a financial rationale for some set of corporate actions related to energy use based on historic and current energy prices. Going the next step to assess what makes sense based on expected future prices increases the challenge of "making the business case." This is particularly true in considering decisions that lock in long-term cost drivers, such as investments in plant and equipment, which must take account of projected market forces and regulatory developments many years in the future. The decision to develop new climate-related products and services, particularly those requiring large-scale, long-term R&D, is even tougher to achieve with quantitative certainty. Assessing the case for any intangible aspects of strategy, such as engagement in policy or protecting reputation, is still more art than science.

Those responsible for building or refining the "business case" can draw several lessons from corporate experience to date—including experience in assessing the rationale for other aspects of corporate strategy under conditions of high uncertainty.

- Begin with a clear understanding of the range of uncertainty in analyzing various strategy elements. Consider creative ways to find numbers to construct the case, but beware of false precision and avoid setting an expectation that the case for most strategy elements will have a clear net present value. It may be helpful from the outset to communicate explicitly to all those involved about what to expect in terms of quantitative and qualitative analysis.

- Frame the initial effort at building a business case as the first step in an iterative process, with the aim of identifying no- and low-regrets strategic options. For example, the initial case can focus on energy-efficiency and clean-energy supply measures that meet the company's investment hurdle rate given existing prices and supply reliability factors. For companies with high exposure to regulatory risk, increased leverage in policy debates may be another obvious point. Certainty about some drivers will increase over time and will allow iterative improvements in the business case, especially if the company makes initial strategic commitments that build internal capacity to understand and respond to climate-related developments.

- It is best to combine "top down" and "bottom up" approaches when building the business case. The top-down approach may be based, at the simplest level, on logic and common sense related to whether the company has a significant financial exposure to the climate issue, whether regulation and other market factors may be coming into play already or soon, whether these conditions call for proactive instead of reactive responses, and which broadly proactive stance fits best with company strategy and culture. The bottom-up approach applies assumptions about market

make money first." More than half of companies surveyed have not been able to quantify bottom-line benefits for their GHG-reduction strategies. Executives warn that this hampers efforts to lock in employee support and create momentum for change. Tim Higgs, Environmental Engineer at Intel, advises against "focusing solely on 'right thing to do' environmental arguments. While this is certainly a key factor in environmental decisions, the case for action is more compelling when combined with more tangible drivers."

E. Making the Business Case for Climate Strategies

How do companies justify GHG strategies if they aren't able to quantify financial benefits? Nearly 50 percent of surveyed companies cite **cost savings from energy-efficiency programs** as evidence of near-term benefits. For example, Calpine estimates it saved $25.8 million over a ten-month period in 2005 by implementing its Plant Optimization Program, which targets thermal efficiency improvements in the company's power-plant operations. Alcoa confirms nearly $80 million in annual savings potential from energy-efficiency improvements and has thus preferences, competitive positioning, regulatory constraints, government incentives, and other drivers to decisions about operations, supply chain management, product development, marketing, and other functional aspects of the company.

- Consider using tools of decision science that help build uncertain and qualitative information into quantitative models. For example, tools for Monte Carlo analysis incorporate uncertainty into financial models in a systematic way. Crystal Ball™ can be used with financial modeling applications to build in ranges of probability that regulation will create a carbon price in any given future year and at any given level.

- Although large companies may have quantitative analyses of brand value, few can accurately predict how consumers, customers, communities, and shareholders will value climate-related factors in the future. Just as climate change and its impacts are not linear, neither is change in public sentiment. Here again, the business case should build in historical examples such as the rapid shift in public attitudes toward clean air in the 1970s and corresponding impacts on company reputations.

- In assessing the business case for policy engagement, look to history for qualitative lessons and even some anecdotal quantitative data. Good examples covered or referenced in this book are DuPont's engagement in domestic and international policy on ozone depletion, Whirlpool's involvement in appliance standards, and Intel's collaboration in advancing the Energy Star™ programs. The financial returns to these efforts are not clear, but it is reasonable to consider these companies' long-term performance and positioning relative to competitors on environmental issues.

- Seek a range of external sources for information and opinions on non-quantitative questions such the shape and timing of climate policies in any given state or country. Industries, particularly industry associations, may face problems of insularity and protracted "group-think" that prevent timely, adaptive thinking about developments in science, business, government, and society. Some good information sources include leading Wall Street firms such as Goldman Sachs and Citigroup and leading consulting firms such as PricewaterhouseCoopers, Mercer, and Deloitte.

- A key tactical tool, which over time can also help build a rigorous business case, is an enterprise-wide energy and environment management system (EMS) that automates operational performance measurement and links this to financial data. Usually safety and health are included as well. The best systems allow managers to look at performance from the plant level all the way up to the entire corporation, and at business unit or geographical "slices" of the company. Best systems can alert managers to poor and excellent performance and allow them to correct problems while also recognizing and disseminating best practices.

far captured annual savings exceeding $20 million. DuPont states that it has achieved an estimated $2 billion in savings since launching its energy-efficiency program in 1990.[51] Given the recent escalation of natural gas prices, similar examples are likely to become more common.

Companies also rely on **less quantifiable methods** to justify their climate-related strategies. One is a general belief among senior leadership that these strategies will add value in the future. "Management believes they add value," says Skiles Boyd, Director of Environment at DTE Energy. "We just haven't been able to quantify it." Some companies believe that getting ahead on this issue offers strategic benefits, such as superior competitive positioning and the ability to identify new market opportunities. For others, getting further out on the learning curve enables them to make the most appropriate investments and prepare to successfully adapt to future regulation. Cinergy treats money spent to reduce GHG emissions, in part, as "tuition to learn." Through such justifications, the business case can be made without precise dollar amounts. As Kevin Leahy, Managing

Director, Climate Policy, explains, "I can't tell you the exact number when it comes to the business case for climate change, but I can tell you the range and an order of magnitude."

Finally, **scenario planning** has helped some companies make the business case for action on climate change. DuPont, for example, has conducted informal scenario exercises that project future business plans and strategies and assess the risk or opportunity implications for families of products. Shell, which has the most experience in this arena, uses scenario planning as a strategic framework for thinking through challenges, identifying risks and opportunities, making investment decisions, developing a common strategic language for leadership teams, and engaging key public policy matters. The most recent (2005) edition of Shell's *Global Scenarios to 2025,* articulates a vision of how worldwide forces might shape markets over the next two decades and reaches the conclusion that the world (and companies) will eventually face a price for carbon. For Shell, this conclusion justifies efforts to increase natural gas production (especially LNG) and investments in wind, solar, biofuels, coal gasification, and experimental hydrogen delivery systems (while still working to make its core business—fossil fuels—succeed in a carbon-constrained world).

F. Other Related Climate Goals and Targets

Some companies have adopted additional climate-related goals and targets. Swiss Re, for example, has committed to increase the renewable share of its energy purchases from 14 percent in 2005 to 37 percent in 2006 and 50 percent in 2007.

DuPont has set three additional climate-related goals as part of its sustainable growth initiative, including a commitment to hold energy consumption to 1990 levels, source 10 percent of that consumption from renewable sources at cost-competitive rates, and receive 25 percent of the company's revenue from non-depletable resources by 2010. So far, energy use has declined by 7 percent compared to 1990 levels, despite a 30 percent increase in production but, as noted previously, the company has only been able to source about 5 percent of its energy from renewable sources, mostly using landfill gas. DuPont is currently two-thirds of the way toward achieving its non-depletable resource goal. This goal represents DuPont's effort toward creating new markets that harmonize with climate constraints. For example, BP and Dupont have targeted 2007 for introduction of biobutanol in the UK as a gasoline bio-component. Once new business opportunities are identified and selected, planning will include goals and targets. As for any business venture, these may include R&D schedules and milestones, product launch dates, revenue targets, market share goals, etc.

In October 2005, Wal-Mart announced the extremely ambitious goal of eventually using 100 percent renewable energy and producing zero waste. These goals tie in with the company's commitment to cut its GHG emissions 20 percent over the next seven years, double the fuel efficiency of its truck fleet within 10 years, and reduce solid waste from U.S. stores by 25 percent in the next three years.[52]

G. Adaptation Strategies

A final, important area of climate-related strategy involves adaptation. According to Ivo Menzinger, Head of Sustainability and Emerging Risk Management for Swiss Re, "No matter what we do now in terms of mitigation, changes in climate are inevitable."[53] Indeed, 60 percent of survey respondents consider physical, climate-related risks to assets in their investment decisions.

The insurance industry is perhaps most directly affected by these types of risks, because it underwrites natural catastrophes and property losses. Swiss Re estimates that total insured property and business interruption losses from natural catastrophes reached $83 billion in 2005. Because climate change directly affects its core business, with or without regulation, the company is integrating related concerns into its underwriting practices, particularly in areas such as Directors & Officers (D&O) and Business Interruption (BI) insurance.

"No matter what we do now in terms of mitigation, changes in climate are inevitable." Indeed, 60 percent of survey respondents consider physical, climate-related risks to assets in their investment decisions.

Companies operating in regions that are especially affected by climate change are also at risk. For example, Diavik Diamond Mines Inc. relies on "ice bridges" to move equipment and materials to the northern regions of Canada. However, the 2006 winter was so warm that roads closed early and the ice never got thick enough to allow transport of the heaviest trucks. The company had to absorb the additional costs of shipping materials by helicopter. In Alaska, the allowable period for traveling on the tundra has shrunk from 220 days in 1970 to about 100 days today.[54] Other impacts of thawing tundra include shifting foundations for pipelines, buildings, and drilling platforms. Further south, warming temperatures can lead to altered growing conditions for agricultural concerns and damage from more extreme hurricanes, such as Katrina and Rita, which impacted oil drilling and refining operations in the Gulf.

In seeking to protect their assets, companies are considering weather concerns as part of their short- and long-term planning and are conducting more extensive resource planning for future plant and market needs. Exelon, for example, expects rising temperatures to alter operational and market forecasts for electricity demand and supply, especially peak consumer demand. This risk is exacerbated by the potential for increased storm severity to damage critical generation, distribution, and transmission systems and produce higher maintenance and capital costs. To prepare, the company is analyzing its ComEd and PECO systems using "worst-case" forecasts of summer peak load based on continually updated information on the likelihood of extreme weather. Exelon has also established emergency preparedness procedures in the event of weather-related disruptions and is planning for increased costs and lead-time to obtain certain supplies. Costs for some supplies could be $30 to $40 million higher in 2006 compared to 2005 due to the after-effects of hurricanes Rita and Katrina.[55]

Physical Assets at Risk from Climate Change

Robert Page, Vice President of Sustainable Development, TransAlta

Climate change is forcing companies to change many financial and asset management practices in order to ensure the viability of existing and future assets. TransAlta, an electricity generator with facilities that stretch from Mexico to Alberta, is now faced with serious issues relating to the sustainability of water resources. Over the past few years water flows for hydro-generation have been unreliable. Some of TransAlta's hydro reservoirs have experienced one year in a century low-water conditions, several years in a row. In the Canadian Rocky Mountains, for example, basic water flows have been affected by receding glaciers and erratic snow packs, reducing the generating capacity of hydroelectric facilities. In the U.S. Southwest there are different water issues: TransAlta chose not to proceed with a thermal power project because the company determined that the available water rights for cooling were not sustainable given climate change and other factors. TransAlta needs 40-year certainty to launch any project. Water sustainability is now a key factor in corporate planning within the company and, as the physical aspects of climate change continue to become more acute, related risks to TransAlta's physical assets and future investments will become more severe.

Gary Serio, Vice President Safety and Environment, Entergy

Climate change poses potential long-term physical risks to the economic vitality of coastal areas. Entergy's franchise territory and assets are particularly susceptible to flooding and hurricanes due to the geographic profile of the territory it serves in the Gulf Coast region. Over the next century, some scientists project average global temperatures will rise five to nine degrees Fahrenheit and sea levels will rise 4 to 35 inches due to projected increases in atmospheric GHG concentrations. Sea-level rise of this magnitude, combined with coastal subsidence, increased hurricane intensity, reduced protection from barrier islands and wetlands losses would exacerbate current vulnerabilities. Physical risks to Entergy include potential damage to power plants, transmission and distribution systems, customer base and other facilities. As an indicator, Entergy's restoration cost for hurricanes Katrina and Rita was $1.5 billion. The social and economic well-being of employees and customers could also be impacted. Entergy has voluntarily reduced its CO_2 emissions 23 percent below 2000 levels over the past five years and is demonstrating economically efficient emission offset transactions. The company also supports mandatory legislation to limit GHG emissions. Sustainable planning for communities should be fostered to adapt to the potential impacts of climate change. Entergy is working with stakeholders for the energy-efficient re-building of New Orleans and the restoration of coastal wetlands. "Hurricanes Katrina and Rita have put a face on the potential risks and financial impacts climate change could mean to our service territory if meaningful action is not taken to reduce atmospheric concentrations of greenhouse gases," according to Jeff Williams, Manager, Corporate Environmental Initiatives at Entergy.

Stage II: Focus Inward

This stage involves integrating climate goals and targets inside the organization by developing supportive financial instruments and engaging employees.

Step 5: Develop Financial Mechanisms to Support Climate Programs.

What are the costs associated with meeting emission-reduction goals and what financial instruments are available for supporting them? This section discusses the pros and cons of internal and external trading, and describes other financial mechanisms used for implementing climate-related initiatives.

A. Lessons Learned

- Costs for climate-related strategies vary widely. Companies can measure these costs along three dimensions: absolute, normalized, and financial-return measures.
- Companies have found internal trading to be of limited value in reducing the cost of actual emission reductions but say it is tremendously useful for educating the workforce and developing expertise.
- Absent legal mandates, U.S. companies are currently using internal pricing mechanisms to support their GHG-reduction efforts, including special pools of capital, lowered internal hurdle rates, and internal shadow prices[56] for carbon. For many companies, the details of these mechanisms are considered proprietary, suggesting that climate strategies are increasingly viewed as a source of competitive advantage.
- Expertise and knowledge gained by developing these mechanisms can help companies understand when climate programs make sense only with an external carbon price and when they can be sustained without one.

B. Cost Estimates for GHG Reductions

On an **absolute-cost** basis, investment in climate strategies varies widely across companies. DuPont knew that the $50 million it spent to develop end-of-pipe controls for N_2O emissions would have no direct payback but decided to pursue the technology anyway, both to preempt government regulation of N_2O emissions and to fulfill its GHG-reduction commitment. In Cinergy's case, the company budgeted $3 million per year in 2004 and 2005 for GHG-reduction projects as part of $21 million the company has set aside for climate programs through the end of the decade. Of the total spent in 2004 and 2005, $4.4 million (73 percent) funded on-system projects and $1.6 million (27 percent) was invested in off-system projects. Resulting annual emissions reductions totaled approximately 600,000 tons and 25,000 tons of CO_2e respectively.

Table 4

Measuring the Cost of GHG Reduction Strategies

Three Examples

Absolute Costs	Normalized Costs	Financial Return
DuPont spent $50 million to develop end-of-pipe control technology to reduce N_2O emissions	Cinergy estimates that the average per-ton cost of CO_2e reductions was $8.28 in 2004 and $12.49 in 2005.	Alcoa has traditionally not pursued climate and energy projects unless they have a payback of one year or less.

Companies also incur costs to develop GHG measurement and tracking systems. Alcoa, for example, estimates that it cost as much as $500,000 to develop its Energy Efficiency Network (after accounting for travel, human capital, and use of internal resources). Richard Notte, Vice President of Energy Services at Alcoa, is quick to add that, "our system is as complicated as anyone is going to get." Whirlpool's call for bids on a data management system to track emissions and conservation yielded cost estimates between $75,000 and $225,000, leading the company to decide to develop a system in-house.

A second way to look at GHG investment is on a **normalized basis**, such as cost per ton of emissions reduced. Again, these numbers can vary widely. Cinergy, for example, estimates that the average cost per ton of CO_2e reductions was $8.28 in 2004 and $12.49 in 2005. These numbers are lower than the $20–$30 range of allowance values seen in the E.U.'s mandatory ETS and higher than the average $2.00–$4.00 per ton values (depending on year) found on the voluntary Chicago Climate Exchange (CCX).[57]

"The most important step is to get all opportunities systematically on the radar screen. Just as every piece of fruit ripens at a different time, not all projects should be pursued immediately. The process starts with quality information."

A third way to consider costs is on a **financial-return basis**. Again, costs using this metric vary across companies. For example, Swiss Re's three-tiered approach for reducing energy consumption is based in part on payback. The first tier is zero cost. The second tier focuses on small investments with paybacks of one year or less. The final tier, which includes refurbishing property, allows for a payback period as long as 10 years. At DuPont, measures implemented by the Titanium Technologies division have led to net year-over-year savings of $3–$5 million. Some projects may have a return of 300–400 percent while others are undertaken with no capital return and are justified on different grounds.

For Alcoa, the availability of capital and the threshold internal rate of return (IRR) required to support GHG initiatives depends on the business situation at individual locations. The company has traditionally not pursued climate and energy projects unless they have a payback of one year or less. As its efforts have matured, Alcoa is moving beyond "low-hanging fruit" investments and implementing projects with longer payback periods. Within its Primary Metals division, energy efficiency projects with an IRR as low as 20 percent are now considered even if the required funds might not be allocated in individual plants' capital budgets. According to Vince Van Son,

Manager of Environmental Finance and Business Development at Alcoa, "The most important step is to get all opportunities systematically on the radar screen. Just as every piece of fruit ripens at a different time, not all projects should be pursued immediately. The process starts with quality information."

C. Internal Carbon Trading

Internal emissions trading has been identified as a highly efficient and accurate way to aggregate information within a company,[58] but of the four companies that have experimented with this tool, none still have programs in place. They all concluded that internal trading did not produce the least-cost, most efficient emissions reductions. Shell, for example, discovered that its STEPS program (Shell Tradable Emissions Permit System) suffered from problems including a lack of participants, a lack of liquidity, and difficulties with permit apportionment. The system was further weakened by the fact that it was voluntary and business units often requested, and received, more permits. Finally and most seriously, there were legal issues: internal emission permits with a monetary value could not be traded across international boundaries without significant tax consequences in host countries.

Despite these difficulties, the STEPS program provided several benefits. It was successful in building awareness among Shell employees, it created a structured mechanism for factoring GHG considerations into the operations of individual business units, it gave the company an opportunity to develop in-house expertise on carbon trading, and it helped the company build credibility in policy circles (Shell's views were considered in the development of the European ETS).

BP claims similar benefits from internal trading. According to Jeff Morgheim, then Climate Change Manager, BP learned "to keep things simple, to get started, to capture the learning and to continuously improve the system. Practical experience is the key to developing a robust system."[59] For Shell and BP, internal trading served as a stepping stone to eventual external trading.

D. External Carbon Trading

External trading programs (like the voluntary CCX or the mandatory U.K. and E.U. ETS systems) offer similar benefits and reduce the need for internal trading systems. According to Interface's Bertolucci, "Interface's participation in external trading programs has helped us to improve our database quality and it has enhanced our ability to track our GHG emissions."

As a founding member of CCX, Baxter International has extensive experience in external trading. According to the company, "Having a goal and reporting on our progress publicly each year has required that Baxter have information systems and verification processes in place to ensure that we are indeed capturing our true performance.... Our emissions in CCX have been entered into a registry and audited for accuracy. This is another learning process for us and because of it we have changed certain things on how we collect and verify global energy usage and calculate associated GHG emissions. This has made our company-wide GHG database more robust.... One of the things we have learned in the auditing process is the need for accurate and easily retrievable energy usage information. This has motivated Baxter to expand the use of external utility payment services, which scrutinize each invoice for accuracy, enter key data into a computer database and scan the actual invoice into the system for possible future reference."[60]

Similarly, Ed Mongan, Global Manager for Energy and Environment at DuPont, says that participation in CCX provided an opportunity to influence the development of trading programs and to demonstrate that market-based approaches are a cost-effective way to achieve GHG reductions. CCX has also helped verify company baseline and annual emissions, which could prove useful for potential revenue generation as DuPont currently has excess emissions- reduction credits. Interface's Bertolucci concurs, "The CCX membership provides valuable third-party validation for what we are doing in regards to climate change."

The precise numbers and formulas companies use for shadow pricing or internal hurdle rates are generally considered proprietary for strategic reasons.

Ultimately, Baxter believes its involvement in CCX will help the company withstand the scrutiny of future emissions verification and trading programs. In the meantime, Meissen sees other benefits: Baxter's involvement in CCX has been widely publicized around the world and has become a source of pride for employees who have been asked to present at various conferences and workshops and have been approached by other companies looking to benchmark emissions or become involved with CCX.

Among surveyed companies, 40 percent participate in voluntary trading programs like CCX, though most note that this participation has not generated revenues. Until mandatory policies create an external market for carbon reductions, companies must develop other means of financially supporting climate projects.

E. Other Financial Instruments

Table 5

Most Common Methods for Funding Climate-Related Investments*

Method	Share
1. A special pool of investment capital	47 percent
2. Shadow prices for carbon	33 percent
3. Lowered internal hurdle rates	32 percent

**Many companies use more than one method.*

Most companies use a combination of approaches to fund their climate-related strategies and evaluate prospective investments. Among those surveyed, the most common methods include reserving a special pool of investment capital; using shadow prices for carbon; and lowering internal hurdle rates (see Table 5).

The precise numbers and formulas companies use for shadow pricing or internal hurdle rates are generally considered proprietary for strategic reasons. For example, Shell uses three different internal shadow prices for carbon: one for the E.U., a second for other developed countries, and a third for the developing world. With these shadow prices, Shell requires that energy efficiency and GHG-reduction projects meet the same internal hurdle rate as other investments. Such internal mechanisms become redundant as mandatory carbon regimes create a real external market price in some locations. By way of illustration, Shell's Hone explains how the value of carbon can be a significant driver in energy-efficiency decisions: One barrel of oil produces about 0.36 tons of CO_2. An E.U. ETS CO_2 price of 25 Euros is like adding a further $11 per barrel to the price of oil, which makes an energy-saving project even more compelling. The company uses long-term premise values for both oil and carbon when valuing internal efficiency projects (the actual numbers used by Shell are confidential and change with the market).

"Our people link our systems and our success. The best technology only gets you so far. Employees will devise innovative ways to achieve clearly stated goals when they understand the linkage with the company's vision and values."

Step 6: Engage the Organization

Employee buy-in is crucial to the success of any climate-related strategy.

As Alcoa's Van Son explains, "Our people link our systems and our success. The best technology only gets you so far. Employees will devise innovative ways to achieve clearly stated goals when they understand the linkage with the company's vision and values." Edan Dionne, Director of Corporate Environmental Affairs at IBM believes that the company's climate strategy "has had a positive impact on recruitment and retention." This section describes techniques for promoting workforce buy-in, identifies common sources of resistance, and describes ways to move climate goals from the periphery of the organization to its core.

A. Lessons Learned

- Given the long-term and complex nature of the climate issue, gaining buy-in from the workforce takes time and effort. Companies find that communication strategies work best if climate initiatives are linked to more familiar issues.
- Many companies link climate-change goals to rewards, bonuses, and public awards. Others employ novel techniques such as promoting tree planting, participation in personal GHG reduction programs, or the purchase and use of low-emission vehicles and bicycles by employees.
- Senior leadership—in the form of speeches, policy statements, Congressional testimony, financial resources, and personal support—is critical.
- Within an organization, it is important to identify the departments or functions that will act as change initiators, implementers, and resistors. Survey respondents identify accounting, finance, and marketing as often less supportive of program implementation than other departments.
- The ultimate goal is to move climate change as an issue from the periphery of the organization to its core. Companies often deploy teams to facilitate this process. To sustain long-term efforts, companies need to maintain a department dedicated to addressing climate issues.

B. Gaining Buy-In

Organizations must have an appreciation for the time it takes to educate the general workforce and management. For example, despite Alcoa's progress, Atkins admits the company would be even further ahead if it had worked on this in "year two, instead of year ten. It takes time to educate 130,000 people."

Educating management can be equally challenging. According to Intel's Higgs, "Climate change is a more difficult subject to convey to management due to the complexity and scope of the issue and the relatively tiny impact of an individual corporation. Other environmental issues are often more acute and therefore easier to drive understanding on why the company should take action."

Companies that have struggled to generate internal support for GHG reductions emphasize the importance of an effective, easily understandable communication strategy. (see "An Energy Efficiency Champion on the Ground" on page 97) "When you talk about trading, impact on energy and economics, you need something besides words. It's hard stuff," says Cinergy's Leahy. Knowing the audience is critical. "You need to ease people into the discussion. Link it to what they already know is possible," says Leahy. "For us, it was our experience with cap-and-trade in our acid-rain program." Whirlpool ties climate change to long-standing company priorities and even refrains from using the term in internal discussions, preferring instead to employ the more familiar terminology of energy efficiency. "We've got a train moving on efficiency," explains Whirlpool's Dahmer. "We'd just start confusing things if we tried to start a new train."

Companies have used traditional and innovative programs to build internal awareness. (see "Ways to Gain Buy-In for Climate-Related Strategies" on this page) DuPont, for example, ties related performance metrics to employee bonuses and has created an award program that recognizes exceptional environmental achievements throughout the company. Rewards and public recognition are common methods of creating buy-in for corporate initiatives.[61]

Alcoa purchases trees from local suppliers and distributes them to employees who are then encouraged to plant them in their communities or on Alcoa property. As of 2005, 1.5 million trees have been planted toward the company's goal of planting 10 million trees by 2020. The company is also encouraging employees to participate in local and regional programs like Smart Trips[62] to increase the use of public transportation and reduce their personal carbon footprint.[63] Swiss Re hosts a wide variety of internal marketing events, including on-site demonstrations that allow employees to test-drive hybrid vehicles. Other companies provide incentives for purchasing hybrid cars.[64]

Ways to Gain Buy-In for Climate-Related Strategies

- Gain support from senior leadership.
- Identify change initiators, implementers, and resistors.
- Develop both cross-functional and specialized teams.
- Create a clear connection between climate change and business strategy.
- Implement specialized internal programs:
  - Tie performance to rewards and bonuses.
  - Create public recognition through award programs.
  - Encourage employees to plant trees to offset emissions.
  - Create internal marketing and educational programs.
  - Encourage participation in programs like Smart Trips and the One-Ton Challenge.
  - Offer financial support for purchasing low-emission vehicles and/or bicycles.
  - Purchase emission offsets.
  - Encourage telecommuting or teleconferencing.

Google offers its full-time U.S.-based employees a $5,000 subsidy toward the purchase of a vehicle with an EPA fuel economy rating of 45 mpg or higher; Integrated Archive Systems offers a $10,000 subsidy. And Swiss Re offers rebates to its employees of up to CHF 5,000 for personal efforts to reduce greenhouse gas emissions including the purchase of hybrid cars or the installation of solar panels or heat pumps on their homes.

Interface associates in North America are offered the option of having trees planted to celebrate their years of service with the company, in lieu of traditional service awards. For example, its "Legacy Award" option sponsors the planting of 80 to 400 trees through an organization called American Forests. Since 2002, more than 3,400 trees have been planted through this program. InterfaceFLOR Commercial Canada provides interest-free bicycle loans to its employees to encourage alternative transportation and emission reductions. And, Interface's Cool Co_2mmute™ program provides associates with an affordable way to "improve their move" by offsetting emissions from commuting and personal travel. Such programs make the climate issue more tangible to people and connect it to their daily lives, while offering examples of how they can make a difference.

C. Senior Leadership

According to companies in this book, senior-level support and engagement are the most critical components of any successful climate strategy. Among survey respondents, CEO leadership was identified as a key driver at all stages of program development and implementation (see Figures 6 and 7). In the words of Alcoa's Atkins, "On a

Figure 6

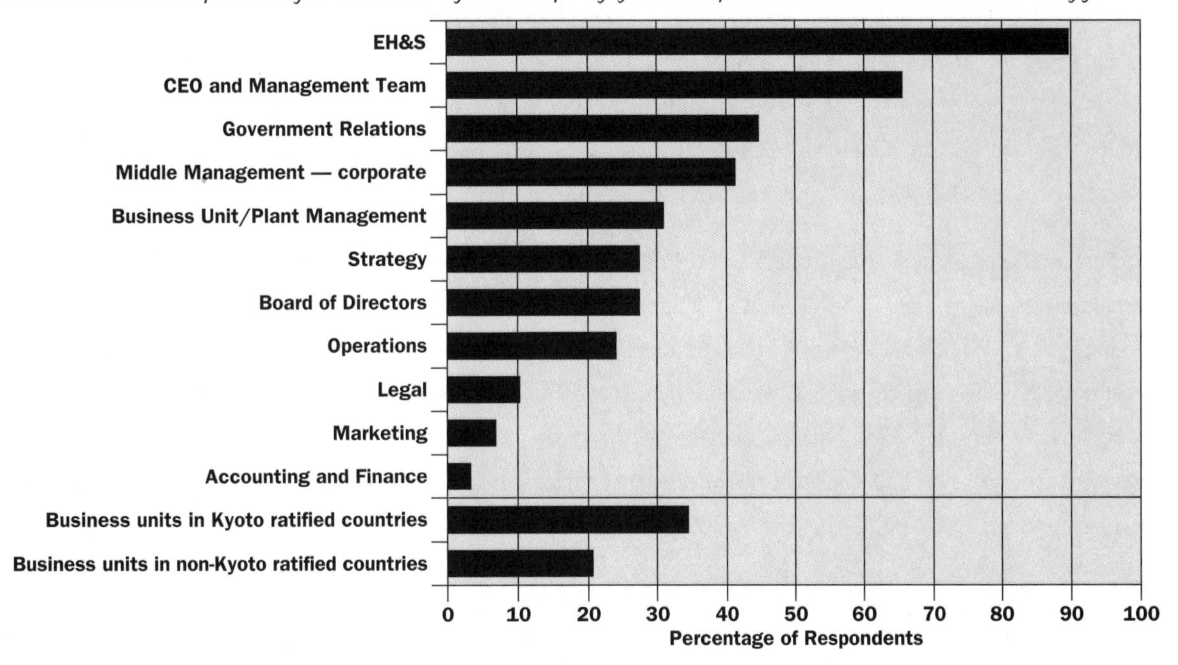

Figure 7

Functions Responsible for **Developing & Adopting** Climate-Related Strategies

Which positions, facets and/or department(s) within your company were significantly involved in developing and adopting your corporate climate-related strategy?

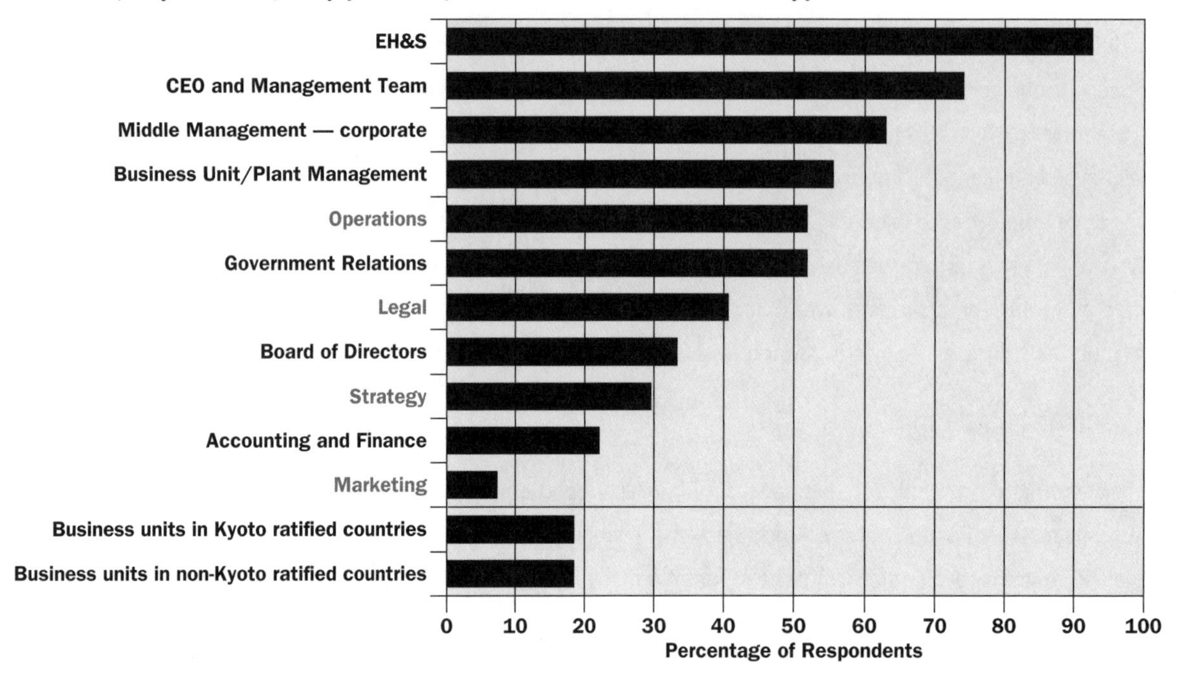

Total Respondents: 27

scale of one to ten, senior-level support is an eleven." When asked about most important lessons learned, Melissa Lavinson, Director of Federal Government Relations at PG&E, notes that, "It is critical to have buy-in at the highest levels and to have the commitment of senior management. It is also important that the Board of Directors understand the business impacts, and opportunities, associated with addressing climate change." Despite the importance of senior-level leadership, CEOs from over 33 percent of surveyed companies have yet to make a public statement on climate change or energy efficiency.

Senior leadership can demonstrate a commitment to addressing climate change in other ways, of course. For example, when business units in DuPont were reluctant to push hard to reach the company's first round of GHG-reduction goals, CEO Chad Holliday stepped in personally to emphasize that failure was unacceptable. His commitment was cited by employees as critical to DuPont's early success. Similarly, Alcoa credits former CEO Paul O'Neill with asking the right questions and challenging engineers to improve the smelting process. Other CEOs, such as Cinergy's Rogers, have been visible spokesmen at Congressional hearings and in the press. More recently, Wal-Mart CEO Lee Scott has received considerable attention for the new, more environmentally sustainable path his company is taking.

CEOs that take a strong leadership position on climate are not afraid to challenge their companies to achieve stretch goals and tend to have a long-term strategic perspective that extends decades beyond their own tenure.

Companies in capital-intensive industries, such as DuPont and Alcoa, even talk in terms of centuries. Says Cinergy's Rogers: "If you are a steward, you make decisions on a longer time horizon, looking beyond your own tenure. When you think of it that way, your view changes. We look 20, 30, 50 years down the road."

In contrast to other companies studied, the criticality of CEO support was not as pronounced with Whirlpool's efforts at emissions reductions. JB Hoyt, Director of Regulatory and State Government Relations, admits that top-down leadership would have been important if the company were starting from scratch, but feels there was no need to push a new mindset given Whirlpool's historic focus on energy efficiency.

D. From Idea to Adoption

When initiating change within a company, climate-related or otherwise, the first questions are: Who will be for it? Who will do it? And who will be against?[65]

Change initiators. The great majority (90 percent) of survey respondents identified their EHS department as an initial champion of climate action (see Figure 6). Sixty-six percent also identified the CEO and the management team. Often EHS supplies the necessary technical expertise while senior management provides the necessary leadership. Initiators are more likely to emerge in business units that are affected by mandatory GHG limits.

Change implementers. In the implementation phase, a wider range of departments and expertise becomes involved, although EHS and senior leadership continue to play critical roles (see Figure 7). As responsibilities spread through the organization, all departments and functions become important, with some departments, such as operations, playing an especially important role in implementing reductions. Here, business units subject to mandatory emissions limits (such as units in countries that are implementing the Kyoto Protocol) are no more likely to be engaged than units elsewhere, suggesting that once a company adopts a corporate climate goal, all parts of the organization get involved.

"If you are a steward, you make decisions on a longer time horizon, looking beyond your own tenure. When you think of it that way, your view changes. We look 20, 30, 50 years down the road."

Change resistors. Survey respondents rank the accounting, finance, and marketing departments as among the least involved in developing and adopting climate programs, while departments responsible for corporate strategy are considered only moderately involved (see Figure 7). These departments are also perceived to be less supportive of implementation than other departments (see Figure 8). Ultimately, breaking down internal resistance is critical to success. Survey respondents identify four main strategies for doing this: establish a clear link between the climate-related strategy and company values, demonstrate clear CEO commitment, create a robust business case for climate-related initiatives, and educate the workforce.

E. Moving Climate Change from the Periphery to the Core

Many case study companies describe how climate change began as an endeavor within EHS but diffused from the periphery to the core and, in the process, became an issue of strategic importance to the company.[66] For this to happen, initiators and implementers must create a clear connection between climate and corporate business strategy that requires all core departments to become involved.[67] Every company approaches this challenge differently. For example, consumer-products companies like Proctor & Gamble are motivated when climate is connected to consumer demand and engages product development, while technology companies like Intel are more likely to address the issue when it is connected to the manufacturing process and engages engineering.[68] At GE, the marketing and finance departments have become heavily involved as a result of the company's $90 million "ecomagination" marketing initiative.[69] Some companies have developed new teams to identify and implement climate-related strategies; such teams may be cross-functional or may have particular expertise and be devoted to a narrow goal.

Figure 8

Organizational Resistance and Buy-In for Climate-Related Strategies

What positions and/or departments within your company are significantly involved in the implementation of your strategy, and what is their level of buy-in or resistance toward your corporate climate-related strategy? (Rank their level of buy-in: 1 = Resist; 3 = Neutral; 5 = Embrace.)

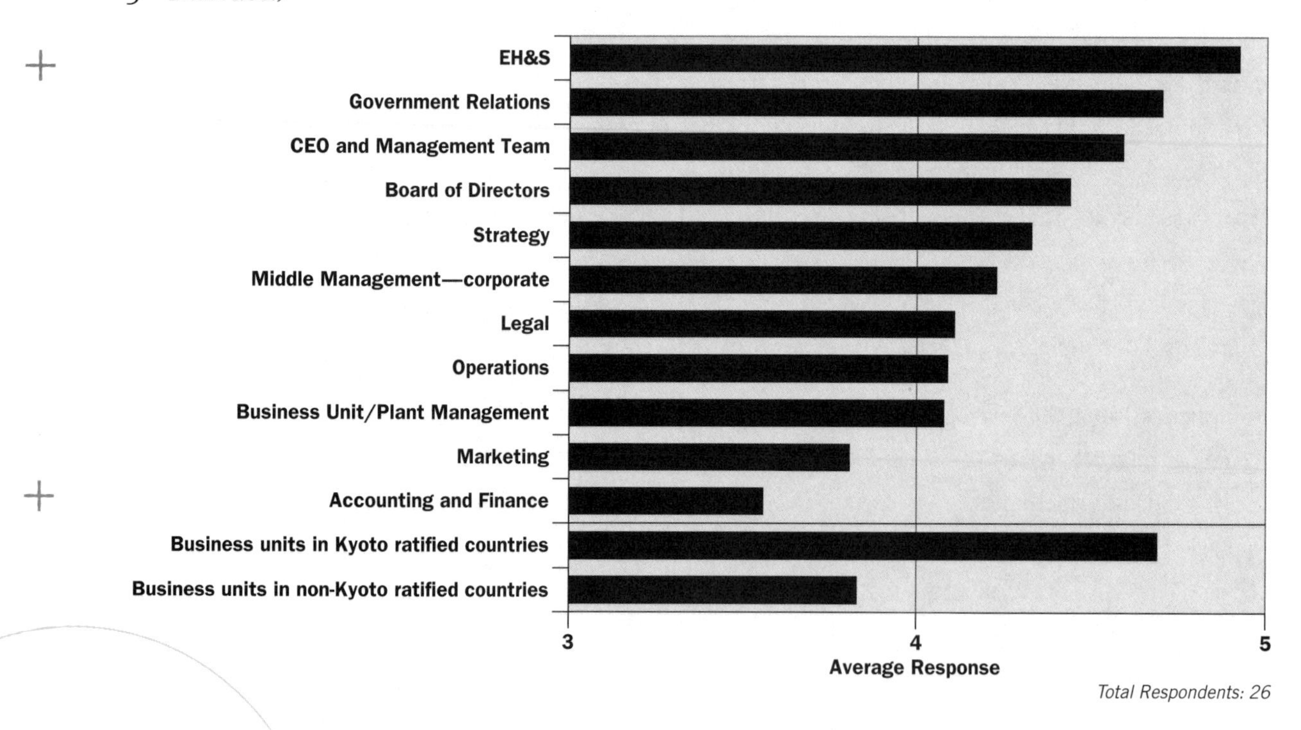

Total Respondents: 26

Market Opportunities from Climate Change

Andrew Casale, Director, Global Marketing and Strategic Planning, Air Products and Chemicals, Inc.

Climate change is driving demand for technologies and solutions that produce cleaner fuels and power. Air Products and its Greenhouse Gases Strategy Team, a cross-functional team led by marketing, is leading efforts to examine the trends, needs, and uncertainties in the regulatory and business environment to create integrated corporate strategies, ensure more sustainable and advantaged businesses, and identify future commercial opportunities. "As a company whose products improve the quality of life for people around the world, every day, as well as help solve the environmental problems of our customers and society at large, we are committed to continuing the journey," says Air Product's CEO, John P. Jones. In an effort to help mitigate emissions from the largest GHG emitters, the transportation and energy sectors, Air Products is actively promoting practical low-carbon technologies for cleaner fuels and power. Work in certain fields, such as the hydrogen economy, carbon capture, and energy efficiency, is an integral part of existing in a carbon-constrained world. Air Products is working with funding and resources from university alliances, venture investments, and the U.S. DOE to develop infrastructure for hydrogen fuels and has already installed over 40 hydrogen fueling stations. Air Products has also demonstrated key technologies in carbon capture programs through in-house R&D, government, and industry partnerships worldwide. Oxyfuel, one example, is a viable option under development for reducing CO_2 emissions, and Air Products has already played a major role in several oxyfuel studies for new-build clean-coal facilities and for retrofitting existing refineries and coal power plants. These and other solutions provide many opportunities for Air Products to continue to provide technology, equipment and services, and products for a sustainable world.

Cross-functional teams: Climate change generally enters the corporate radar screen through existing teams or departments that watch for issues critical to the company's future. Whirlpool, for example, first began attending to climate change in the same way it addresses other environmental issues: through the company's Environmental Council, a group comprised of representatives from its six geographically dispersed business units. Similarly, Interface's Global Sustainability Council is a cross-functional team that looks at climate change and other pertinent issues from a wide variety of perspectives including product development, life cycle assessments, business development, public relations, sustainable operations and reporting, and EHS.

Teams with Focused Expertise: Once on the agenda, companies often develop new teams to focus on climate strategies. For example, Alcoa launched a Corporate Climate Change Strategy Team directed by top executives and comprised of 11 members representing operations, government affairs, technology, communications, and finance and with geographic representation from the United States, Canada, Australia, Europe, and Brazil. According to Randy Overbey, President of Primary Metals Development and the current director, the secret to the team's success is its multi-functional membership: "The members may not always agree with each other, but having such diverse representation increases the robustness of our results."

Cinergy developed a GHG Management Committee to oversee the allocation of its $21 million GHG fund. The committee includes ten senior representatives from business areas that would be affected by GHG policy and one ex-officio NGO member, Environmental Defense. Similarly, Shell has created a new unit, led by senior executive Graeme Sweeney, who is also head of Hydrogen and Renewables, to kick-start and foster GHG-reduction technologies until they are sufficiently integrated in the company's business units to stand on their own.

"GHG is becoming more and more internalized. While we are still learning, it is clear that climate change has to be imbedded in the real business strategy early on and not just remain a Health, Safety, and Environment issue."

Many companies also have groups that explicitly look for energy-efficiency opportunities: an example is Alcoa's Energy Efficiency Network (EEN). DuPont has a similarly purposed Energy Competence Center, while Shell has the Energise group within its Global Solutions internal consulting arm. Each team is slightly different in structure, but all include technical experts drawn from both corporate and local-business-unit levels. Alcoa's EEN augments internal personnel with external experts. In each case, these groups deploy teams at the request of unit managers and perform audits to recommend operational, equipment and behavioral changes (the decision to implement is typically left to site managers). They also identify, document, and disseminate information about successful energy practices observed at plant locations.

The Ongoing Need for Specialized Expertise: Even after climate is integrated in core functions, the need remains for a smaller but dedicated department to identify future business opportunities. At Shell for example, company-wide internal trading began with the Health, Safety & Environment (HSE) group within Corporate Affairs. It was then moved to Shell Trading with the creation of a CO_2 trading desk to allow the company to participate in the Danish and U.K. ETS's. "GHG is becoming more and more internalized," states Shell's Hone, adding, "While we are still learning, it is clear that climate change has to be imbedded in the real business strategy early on and not just remain an HSE issue."

A similar process occurred at Swiss Re, which created a Greenhouse Gas Risk Solutions (GHGRS) department. The group was dissolved in the summer of 2005 and its mature offerings, including carbon trading, insurance products, and weather derivatives, were redistributed to mainline product groups. A centralized logistics department was created to oversee office-space management and carbon neutrality; and Walker, the head of GHGRS, was reassigned as a manager of Sustainability Business Development, which focuses on bringing products related to climate and sustainability to market. By successfully integrating its climate activities with its various mainline businesses, such as Capital Markets and Advisory (trading products), Risk Awareness (D&O insurance) and Carbon/Clean Energy Asset Management, Swiss Re can more effectively engage climate change as a strategic bottom-line issue going forward.

Wal-Mart Mini Case Study[70]

In October of 2005, Wal-Mart CEO Lee Scott announced a series of sustainability goals, of which climate-related goals were central, of remarkable scale and ambition. Wal-Mart, Scott said, was going to buy 100 percent of its power from renewable sources; produce zero waste; double the fuel efficiency of its trucks; avoid greenhouse gas emissions by 20 percent; and challenge thousands of its suppliers to follow its lead. "People expect a lot of us, and they have a right to," Scott said.[71] "Due to our size and scope, we are uniquely positioned to have great success and impact in the world, perhaps like no company before us."

Size and Scope

The words "size" and "scope" are key to understanding Wal-Mart's environmental impact and its sustainability strategy. Wal-Mart itself is huge, but its scope is almost incomprehensibly vast, with a supply chain estimated to be made up of between 30,000 and 60,000 companies. Wal-Mart organized its sustainability strategy under three main goals, two of which—increasing renewable energy purchases and cutting down on waste—the company characterizes as direct goals. The third goal, developing sustainable products, is an indirect goal, which Wal-Mart will work with its suppliers to achieve. According to Andy Ruben, Wal-Mart's vice president for corporate strategy and sustainability, the indirect goal was recognized early in the process by company management as the real prize of the sustainability strategy. "We knew that 90% of our ability to create change was through our supply chain," he said. For example, Wal-Mart recently calculated that its direct greenhouse gas emissions stand at a little over 20 million metric tons (MMT) of CO_2 equivalents. But it estimates that emissions from all of its suppliers could top 200 MMT. This does not include emissions from the use of the company's products—appliances, electronics, light bulbs, etc.—which very rough estimates place in the hundreds of millions of tons of CO_2 equivalents, Ruben said. The real success of Wal-Mart's sustainability strategy, therefore, will be measured by the extent to which the company is able to influence and support the behavior of its suppliers and customer base. "20 million metric tons is a scratch compared to what we can really change," Ruben said. "But in order to get there, we need to walk the walk." In other words, Wal-Mart's direct goals, while significant in their own right, are also designed to send a message that the company is serious about its sustainability initiative, and give it the credibility to demand change from its suppliers.

From Defense to Offense

But while the goals are ambitious and aggressive, Wal-Mart executives freely admit that the motivation behind the company's sustainability strategy was initially "defensive." "We started thinking, if we could go back 10 years, what are some of the things we would have done to avoid some of the issues we're facing now," said Ruben. Then the company looked 10 years ahead and tried to imagine some of the things it might be criticized for in the future. "And the environment was one of those things," Ruben said.

What began as a defensive strategy, however, soon turned into something much more proactive as company executives and rank-and-file associates embraced the initiative wholeheartedly. "No one here imagined we'd be where we are a year and a half ago," Ruben said. The key to the transition, according to Ruben, was having support from the highest levels of the company. "There's nothing that can compensate for not having top-level support." Ruben describes CEO Lee Scott as being "100 percent engaged" in the strategy. But Scott's approach has not been to issue top-down mandates. Instead he has focused on persuading staff of the merits of the strategy and convincing them to pursue it based on its business benefits. "He's actually changing the culture of the company," Ruben said.

Crafting the Strategy

The development of the strategy began in June of 2004 when Scott first met with and engaged Jib Ellison from Blu Skye consulting. One of the first things Ellison and two other consultants from Blu Skye did after holding their initial meeting with Scott was develop a rough "back of the envelope" estimate of Wal-Mart's environmental footprint. That analysis made it clear that the company's main environmental impact resided in its supply chain. Following completion of the analysis, from June through September Ellison and his team conducted a series of hour-long interviews with top-level executives from Wal-Mart to get their thoughts on the nascent sustainability initiative.

The next major step came in September, when Ellison organized what he calls a "choice meeting." The meeting was specifically designed to put the sustainability initiative "to a choice," and give Wal-Mart officials the opportunity to opt out if they wanted, Ellison said. The invite list for the meeting was carefully crafted. CEO Scott was there, along with about 25 other top Wal-Mart executives. In addition, around 30 of the company's rising stars were invited. "We wanted the future rock stars of the company," Ellison said. With the exception of the head of Wal-Mart's environmental, health and safety (EHS) department, Ellison purposefully chose not to invite any of the company's Associates (employees in Wal-Mart are called Associates) who had existing environmental responsibilities. The decision to exclude Wal-Mart's environmental personnel was designed to advance the principle that sustainability should be fully built into the business and not viewed as an add-on or extra component. Ellison believes that many consultants and nongovernmental organizations that advise on climate and sustainability mistakenly focus their efforts on lobbying EHS officials, which helps foster the misperception that sustainability initiatives should be pursued separately and distinctly from the company's general business strategy. "My insight was no, this is a business strategy...the whole key is to switch this from a burden and a duty to an opportunity and something I want to do," he said. By getting Wal-Mart's rising business leaders involved in the process early, Ellison was better able to convey the idea that climate action and sustainability is a serious business growth opportunity and not an additional burden.

The September meeting spanned two days, with the first devoted mainly to educating the attendees on environmental sustainability in general, why it is a business issue, how it is affecting Wal-Mart, and how it can be used to create shareholder value. The consultants also described the company's environmental footprint. Then presentations were made by a couple of Wal-Mart suppliers who have already embraced sustainability and made serious commitments in that area. Presentations were also made by representatives from Conservation International. On day two, after Ellison felt the meeting participants were well enough educated on the company's footprint, sustainability in general, and the business risks and opportunities involved, he presented them with a choice: "Now what do you want to do?" The consultants left the room to let the employees discuss the issue privately among themselves. When the consultants came back, the Wal-Mart employees told them that they were unanimously in support of pursuing a strategy. The rest of the second day was devoted to thinking about how to organize the development of the strategy.

It was also in September that the initiative was first discussed with Wal-Mart's board of directors. According to Ruben, the board responded very positively to Wal-Mart's emerging sustainability strategy and has continued to be supportive. "They've been urging us for years to take a more external view...to think about what the world looks like outside of Bentonville, Arkansas," he said. Still, the board has not taken a hands on role in the development of the strategy.

In December 2004, Wal-Mart held a follow-up meeting with most of the same attendees from the September meeting. The winter meeting focused on how to engage with NGOs, particularly those critical of Wal-Mart. Ellison wanted Wal-Mart to learn how to deal constructively with its critics, because, "this is a company that's been extremely isolated, in a lot of respects." The December meeting became the model for the tri-annual meetings that are now known as "Milestone Meetings."

The Milestone Meetings have three basic components: education; progress reporting; and forward planning. The education component is important so that Wal-Mart employees continue to learn new things as they move forward with the initiative. The progress reporting element is designed so that everyone can stay up-to-date with actions taken on the sustainability strategy. Ellison also considers the progress reporting element to be a crucial accountability mechanism. Employees know that at these meetings they may get called up in front of Lee Scott or board chairman Rob Walton to say, "I got it done, or I didn't get it done." They keep that in the front of their minds leading up to the meetings, which pushes them to make real progress on their climate and sustainability responsibilities. The "forward planning" component is essentially a way for the company to talk about and organize future actions. Additionally, Wal-Mart is bringing in more outside parties to the meetings for educational purposes, and as a way to build relationships.

Organization and Implementation

Wal-Mart organizes the elaboration and implementation of its sustainability strategy through 14 "sustainable value networks," which grew out of four initial "clusters" that company officials organized themselves into at the December meeting. Those clusters dealt with business communication (internal and external); products and supply chain; energy efficiency; and waste. Those clusters grew into the 14 networks that now cover the following topics: global greenhouse gas strategy; alternative fuels; energy design construction and maintenance; global logistics; operations and internal procurement; packaging; textiles; electronics; food and agriculture; forest and paper; chemical intensive products; jewelry; seafood; and the China sustainable network. Each network is led by a Wal-Mart employee and will often include five or six suppliers—including both direct and indirect suppliers—as well as outside experts invited from nongovernmental organizations, academia, and the government.

Organizing the climate and other sustainability strategy workstreams through 14 relatively decentralized networks has worked well for Wal-Mart, though it may not work for all companies, Ruben said. There are certain tradeoffs involved, and one of the disadvantages is that not all employees are equally motivated, and some networks move faster and more aggressively than others. Conceding a level of frustration that not everyone is moving at the same pace, Ruben said the company's promotion process is one way of motivating employees. "One thing that happens is that the people who are being promoted faster tend to be the ones who are involved in this sort of thing.... That gets the message across faster than any memo you can send."

Ruben also believes that employees will be more motivated if they perceive a clear business case behind the sustainability strategy. It has helped that the company organizes its initiatives into three buckets: "quick wins," which include things that are relatively easy to accomplish and provide almost immediate returns, like cutting waste and improving energy efficiency; "innovation projects," which are longer-term initiatives with less immediate payback, like the company's plans to sell some emissions credits from its supply chain reductions into carbon markets, and then reinvest the proceeds into GHG-reducing projects; and "game changers," which have the potential to fundamentally alter the way goods are bought, sold, and used. "What's been important is to take all the quick wins to fuel your growth," while continuing to go after the innovation projects and game changers at the same time, said Ruben.

Only five people work full-time on the sustainability strategy in a company that employs about 1.8 million people overall. And all of the heads of the "sustainable value networks" are doing that work in addition to their previous primary responsibilities, which has led to some concerns that the initiative is under-staffed. But Ruben believes Wal-Mart has the staffing levels set up in a way that is consistent with Ellison's philosophy that the sustainability strategy must be incorporated deep into the core of the business and integrated throughout the company. "It forces you to think about how this can live in the business." And the staffing levels are also consistent with Wal-Mart's general approach, he said: "Look for constraints that make you better."

Stage III: Focus Outward

This stage of climate-strategy development involves engaging important external constituencies that directly impact strategic success. To have external legitimacy, companies first need to establish a track record of credible internal action.

Step 7: Formulate a Policy Strategy

Companies must consider how different external GHG policies can affect their business objectives. At the most basic level, this means monitoring and anticipating pending government actions. Beyond that, companies must be aware of the policy options being considered and decide which would most benefit their own business strategy. At the highest level, companies will want to gain (and maintain) a seat at the table when future regulations are designed.

A. Lessons Learned

- All companies acknowledge the tight link between government policy and business strategy and see a strong need to participate in policy development. This participation can be geared toward advancing individual-company interests, but it is also seen as way of providing valuable input toward the goal of sound and effective policy.
- Nearly all companies (90 percent) in this book believe that government regulation is imminent, and 67 percent believe it will take effect between 2010 and 2015, a timeframe that is consistent with the 2010 start date of emission trading systems proposed by the NCEP[72] and the McCain-Lieberman Climate Stewardship Act.
- There is broad agreement among companies about several key aspects of prospective policy: market-based trading, sequestration credit, the need for federal regulation to supersede a growing "patchwork quilt" of state regulations, and credit for early action. Companies differ, however, in their views on other issues, such as the baseline date for reductions, credit for (and definition of) indirect emissions, and preference for sector-based versus economy-wide policy.

B. The Link between Policy and Strategy

All case study companies acknowledge the strategic value of having a seat at the table to influence policy development. In fact, most have a long history of working with government on environmental policy. Cinergy's Rogers feels that involvement with government is necessary to avoid "stroke of the pen risk, the risk that a regulator or Congressman signing a law can change the value of our assets overnight." Rogers continues, "If there is a high probability that there will be regulation, you try to position yourself to influence the outcome." About the development of cap-and-trade programs, Shell's Hone says plainly, "If you're doing a deal with somebody and they're setting the rules, then you want to have a say."

While companies consider it a business opportunity to advocate their desired policy, some also believe their participation is necessary for good policy. According to DuPont's Fisher, "It is important for industry to help government find cost-effective solutions to the climate change issue. Government can't do it alone. They don't have the capacity to understand all the implications of the different policy options." Carolyn Green, Vice President of Health, Environment and Safety at Sunoco, goes further, citing "how little environmental regulators and advocates know about the energy intensity of their requirements."

Involvement with government is necessary to avoid "stroke of the pen risk, the risk that a regulator or Congressman signing a law can change the value of our assets overnight."

Other reasons for engaging in the policy-development process include: lobbying for subsidies or other support for strategic initiatives (such as IGCC), pre-empting regulation by demonstrating that action is already underway, deflecting policy toward other firms or industries, or convincing regulators that a policy will not be costly and should therefore be imposed on the entire industry.[73]

C. Policy is on the Horizon

Despite little progress toward national GHG regulations, all survey respondents believe that government involvement is necessary to address climate change. According to Exelon's Pagano, "We believe that leading companies will do what they can do in advance of mandatory programs, but we believe that to go beyond the base level of effort that is occurring in the voluntary period and to make significant progress in addressing this global issue, government mandates will be required." Cinergy's Leahy adds, "The technologies will emerge when CO_2 has a price signal, and that market signal will be created by regulation."

Companies have also begun encouraging action in the global arena. Similar to the findings of a 2004 Pew Center report (*International Climate Efforts Beyond 2012: A Survey of Approaches),* many case study companies draw connections between U.S. action, actions by other nations, and the economics of international carbon markets.[74] According to Michael Parr, Senior Manager of Government Affairs at DuPont, "We won't see China and India on board while the U.S. is on the sidelines." As a result, "Market liquidity of carbon credits is restrained without a global market," explains David Rurak, Director of Operations at DuPont.

"The technologies will emerge when CO_2 has a price signal, and that market signal will be created by regulation."

D. Options for Policy Mechanisms

Notwithstanding the wide range of industry sectors represented, the survey revealed broad agreement in a number of key policy areas (see Figure 9). At the top of the list is GHG trading. Research recently conducted by Deloitte supports the importance of this issue: "Trading in emission permits will enable power and utility companies to stay within the rules even though they may have difficulty cutting their emissions rapidly due to technology gaps and cost issues."[75] Baxter's Meissen envisions a future "not only with active regional and national GHG trading markets but also with interconnected markets among established financial centers of the world." With trading, Cinergy believes that climate policy can be instituted in a manner that avoids significant costs to the economy. "What is important is that lawmakers know that even some coal-fired utilities think it is possible to deal with the climate problem without harming the economy," says Cinergy's Leahy.

Figure 9

Anticipated Features of Future Climate Change Standards

What kinds of actions will be most important [in federal standards on climate change]? (Please rate their level of importance: 1 = not important; 3 = neutral; 5 = important).

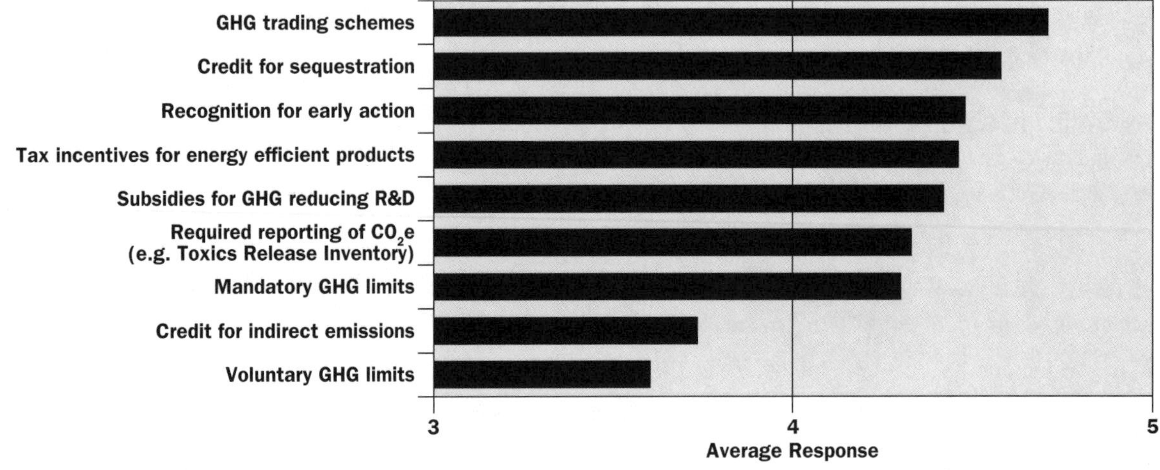

Total Respondents: 26

A second priority for many companies is ensuring that future policy allows credit for biological and geological carbon sequestration. The prominence of the latter is notable because it was ranked lowest in terms of providing bottom-line benefits to companies absent a carbon price signal (see Figure 5). Given how critical sequestration may be to the future of coal, support for carbon capture and geological sequestration (through subsidies or research and development support) is of paramount importance, especially to electric utilities.

A third priority—especially for early adopters who have already exploited "low-hanging fruit" opportunities—is receiving credit for previous emissions reductions. For some companies, such as DuPont and Alcoa, it is *the* critical issue. According to Jake Siewert, Alcoa's Vice President of Environment, Health Safety, Global Communications

Carbon Capture and Storage is a Critical Need

Marcelle Shoop, Director Environmental Policy and Partnerships, Rio Tinto

Rio Tinto recognizes that the long-term future use of fossil fuels, including coal, depends on widespread deployment of low emissions technologies such as carbon capture and storage (CCS). "The challenge for this century is to reduce CO_2 emissions from fossil fuels such as coal. We must use clean coal technologies, but the key to unlocking an environmentally friendly future for fossil fuels is secure carbon storage," says Preston Chiaro, Energy Group Executive for Rio Tinto. However, the widespread application, public acceptance, rapid commercialization, and ultimate success of CCS will depend on:

- Identifying geologic sequestration potential, including potential in less explored areas;
- Improving the understanding of the permanence of CO_2 storage by developing appropriate techniques for monitoring, measuring, mitigating, and verifying the effectiveness of long-term CO_2 storage;
- Reducing CCS cost through significant public and private investment in research, development, demonstration, and deployment of various forms of capture and storage technology, but particularly at a large scale;
- Addressing key legal and policy considerations associated with the deployment of CCS technology, including issues of ownership and liability; regulatory policies for measuring, monitoring, and verification; and siting.

Rio Tinto is actively working with industry associations and international government-industry partnerships such as the Carbon Sequestration Leadership Forum and the Global Energy Technology Strategy Program to support key R&D activities including CCS technology and exploring CCS regulatory frameworks. In the United States, Rio Tinto is a founding member of the FutureGen Industrial Alliance, which is developing clean coal technology, including CCS. In Australia, Rio Tinto is a founding contributor to the recently announced Coal21 Fund, a voluntary coal-industry levy which aims to fund CCS projects. Through strong leadership and careful investment in research and demonstration, Rio Tinto and others will play an important role in developing and deploying emissions-reducing technologies like CCS.[76]

and Public Strategy, "Although I can't imagine anything coming out of Washington that would be too strict for us, the worst-case scenario is not getting credit for what we've already done and having to start today."

To be positioned to gain credit for early action, companies have been careful to register their reductions through a variety of mechanisms. Cinergy, for example, reports its emissions reductions through the DOE's EIA 1605(b) reporting system and to EPA as part of the Climate Leaders program. Swiss Re plans to register its emissions reductions with the World Economic Forum's GHG registry. Whichever mechanism is used, this is a critical step for early adopters.

Companies' preferred baseline date for purposes of future regulation usually corresponds with the date when they started reducing emissions. The median answer to this survey question was 1990 and the average date was

1994, consistent both with the 1990 baseline set by the Kyoto treaty and reflective of when most respondents began taking action. These companies' primary concern, irrespective of what date is chosen, is credibly certifying early reductions.

Many companies agree with Entergy's Williams, who believes that "Policies need to allow price signals to be sent that will allow flexible investments in energy efficiency and clean, non-emitting generation technologies, such as renewables, nuclear, IGCC coal with carbon capture and storage. These investments will help keep the cost of a mandatory program low." A few companies also note the need for unified federal regulation to supersede a patchwork of state and local actions, which they believe place an unnecessary burden on manufacturers. In the words of Tom Catania, Vice President of Government Relations at Whirlpool, "This would be a huge misdirection of resources and much less would be achieved if we are subjected to a balkanized set of standards from 50 different sources."

In other key policy areas, company positions differ. For example, some companies, such as Holcim, prefer sector-level emissions caps because they are concerned that one sector (such as transportation) might otherwise bid carbon prices to a level high enough to adversely impact another sector (such as manufacturing). Some have suggested that a sector-specific approach would prevent energy-intensive industries, which are seen to have the most at stake, from capturing the regulatory process.[77]

"This would be a huge misdirection of resources and much less would be achieved if we are subjected to a balkanized set of standards from 50 different sources."

Other companies favor economy-wide approaches that cover all industries under one cap. In a recent white paper, Duke Energy stated: "Exclusions of sectors or GHGs from a program would be unfair and economically inefficient, and would reduce program effectiveness." The white paper goes on to recommend that "the point of regulation should be upstream...Downstream and other approaches would likely result in more limited coverage, fragmented program approaches, economic inefficiencies and greater administrative complexity and costs."[78]

Companies also differ on how they want policy to treat indirect emissions. For example, a number of manufacturing companies want credit for emissions reductions related to the use of their products. Maytag and Whirlpool consider use-phase reductions so important to their respective strategies that without it their involvement in carbon markets will be seriously limited. Other manufacturing companies, such as Alcoa, are less concerned about in-use emissions and are more interested in how GHG regulations will affect market demand for their products. Alcoa anticipates increased sales from, for example, continued light-weighting of automobiles.

In the end, all companies recognize that establishing climate policy will be a challenge, despite their universal belief that such policy is needed. Cinergy's Leahy believes it will be very difficult to justify GHG regulations to the average voter: "Advocates for a carbon control regime should be prepared for an aggressive media campaign by opponents—who was that couple we saw in the early 90's during the health care debate? As soon as anything looks like it may become law, we'll see them again, only this time they'll say, 'Honey, did you know we're going to get hit with an X percent tax on energy use?' 'Wow, that's going to force the price of everything UP!' 'Yes, and it says here X hundred thousand people will lose their jobs because of this!' It's tough to fit an accurate picture into nice sound bites, especially for such a complex issue." Just as much of the opposition to acid rain legislation was based on projected costs that were much higher than the ultimate reality,[79] similar cost concerns are likely to present a major hurdle to climate regulation.

Step 8: Manage External Relations

One final component of a successful climate strategy is engaging external constituents including competitors, trade associations, suppliers, customers, regulators, and NGOs. All case study interviewees note that these groups provide vital information and expertise, can help develop markets and support for climate-related initiatives, and are important adjudicators of credibility and reputation. As described in this section, firms must identify critical target audiences and understand their connection to company objectives.

A. Lessons Learned

- External outreach is critical to success. Outside groups can provide knowledge and key avenues for advancing business objectives.
- The external outreach efforts of survey respondents were aimed first at employees and NGOs, and then at government, the broader public, and the investment community.
- External groups sometimes oppose climate initiatives. Government and trade associations were named as the number one and two sources of resistance. In response, all companies engage in some form of federal- or state-level lobbying, and many work within trade groups to create change.
- Survey respondents also report reaching out to customers and other companies through research consortia, trade groups, and other avenues.
- In the end, all the steps in a firm's climate strategy have to fit with each other and with overall strategic objectives. External perception must not be different than internal reality.

B. Target Audience

All of the case study companies engage in external outreach as part of their climate strategy, most commonly to promote transparency and stakeholder dialogue. But at whom should these efforts be directed? Identifying the target audience is pivotal to successful outreach. Public reporting in one case may be strategic communications in another.

According to Mirza, Holcim reports information publicly "to establish to our employees, the communities in which we operate, customers, investors, and governments that we recognize this as a significant environmental aspect of our operations, and that we are taking action to address it." For Interface, Bertolucci believes the company's public outreach strategy has helped it become "internationally recognized as a sustainability leader." At Shell, the company's annual *Sustainability Report* serves three purposes: to present the company's public face and report its activities to the outside world, to give staff and different business units a guiding vision, and to allow those units to communicate concerns and ideas during the process of compiling the *Report*.

The survey results reflect a heavy emphasis on internal audiences. As noted previously, respondents state that external outreach efforts are aimed first at employees (a somewhat counter-intuitive finding) and NGOs, followed by government, the broader public, and investors (see Figure 10). Each represents a different audience and requires a different form of outreach.

Figure 10

Targets of **Public Reporting and Communications**

How important are the following groups to your company in communicating about its climate-related strategy? (Rate their level of importance: 1 = not important; 3 = neutral; 5 = important).

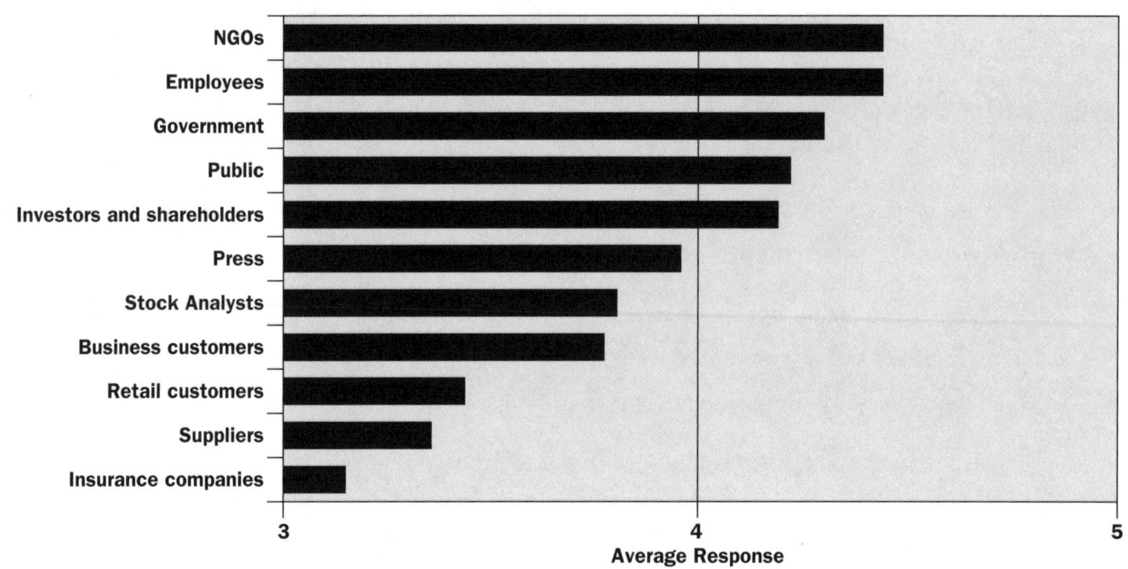

Total Respondents: 27

Outreach to **employees** is discussed in previous sections that cover the need for workforce buy-in and internal support (see Step 6: Engage the Organization on page 37).

All companies in this book have worked with **NGOs**, such as Environmental Defense, the World Resources Institute, and the Pew Center. Benefits from such engagement include access to scientific, technical, or policy information; access to specialized expertise; information sharing with other companies (through NGO-created consortiums like the Green Power Market Development Group and the BELC); the opportunity to test ideas before releasing them to the public; or verification of corporate emission reductions. NGOs can work with companies through formal working groups, internal panels or boards, corporation-funded initiatives to promote technical research or public awareness, and as representatives of local community interests.

Many companies have found that NGOs can be tremendously helpful. According to DuPont's Fisher, "You can learn a lot from NGOs. They can open your eyes to market opportunities. Also, they add legitimacy to our environmental commitments. A big, branded corporation stating its efforts sounds like public relations, but an NGO recognizing them carries a lot of weight, both internally for employees who are passionate on the subject and externally." Alcoa's Siewert concurs: "We know we're not the expert on these issues; we need help. Our people broaden their view of sustainability by interacting with others who think more broadly, with the people who help manage the growth process more effectively. When we think too narrowly, we get in trouble because the rest of the world doesn't think that way."

Shell, for example, worked with a panel of NGO and Native American tribal representatives as part of its Canadian Athabascan oil sands project. When Cinergy first developed a baseline emissions assessment, it partnered with Environmental Defense to help validate the process. Environmental Defense reviewed Cinergy's definition of its corporate emissions footprint, approved the methods used to identify and measure GHG reductions, evaluated the company's implementation of a GHG fund, and now serves as an ex-officio member of Cinergy's GHG Management Committee.

"You can learn a lot from NGOs. They can open your eyes to market opportunities. Also, they add legitimacy to our environmental commitments."

Alliances with NGOs also provide credibility for both parties. Shell's work with the Pew Center, for example, opens doors. "Once you go through Pew," Hone says, "it's like you've gone through a filtering process—you have additional credibility. Shell provides Pew with credibility. And likewise, Shell gets the same. There is less suspicion than if Shell went it alone." Whirlpool has used similar alliances to further mutual policy interests; for example, the company worked closely with the Sierra Club, Natural Resources Defense Council, and the Alliance to Save Energy to promote manufacturer tax credits for efficient appliances in recently adopted energy legislation.

Governments are targets of external outreach when companies seek to gain insights about—or have an influence over—the likely form of future regulations (see Step 7: Formulate a Policy Strategy on page 48).

The **public** is also an important audience for external outreach on climate issues. Swiss Re is the world's largest re-insurer and, more than any other company in the book, is concerned about general societal awareness. The company has therefore engaged in a broadly focused external outreach effort using some of the more unorthodox techniques documented in this book. For example, the company sponsored a documentary called *The Great Warming* that was broadcast in 2005 on the U.S. Public Broadcasting System (PBS). It also partnered with the United Nations Development Program and Harvard Medical School to host a conference and produce a report called *Climate Change Futures: Health, Ecological and Economic Dimensions.*

Enhanced Financial Disclosure of Climate Risk and Opportunity

Yolanda Pagano, Director Climate Strategy and Programs, Exelon Corporation

Despite having submitted GHG information through the U.S. DOE's 1605(b) program and to the U.S. EPA pursuant to Title IV of the 1990 Clean Air Act Amendments, and having published CO_2 emissions in its annual environmental report, Exelon sought to disclose more. Helen Howes, Vice President, Environment, Health & Safety, states that, "Our shareholders wanted to better understand the opportunities and risks that the climate change issue represented to their investment in Exelon, so we added a Global Climate Change section to our 2004 10-K."[80] This decision was influenced by Exelon's commitment to develop a voluntary greenhouse gas emission reduction goal through the U.S. EPA's Climate Leaders Partnership; heightened interest in climate change with the ratification of the Kyoto Protocol by Russia in 2004; and the rise in investor interest as evidenced by both the increasing number of shareholder resolutions at corporations and the direct requests from investor groups such as the Carbon Disclosure Project. Exelon feels secure in its competitive position given its large fleet of low-cost, non-carbon emitting nuclear generating assets and has disclosed its position in support of the science on climate change and the need to take action now to deal with climate risks.

All public companies pay close attention to the **investment community**. Some companies feel their climate strategies have kept investors from filing proxy resolutions or exerting other forms of pressure. Other companies actively engage investors on these issues. Alcoa for example, has convened meetings with its top investors to discuss sustainability concerns. Survey respondents and case-study interviewees note that interest until quite recently has been limited to socially-responsible investors. But they anticipate that mainstream investors may play a larger role in the future. "The mainstream investors are not as strong on this issue in the United States as they might be, but that could all change if legislation is enacted," says DuPont's Fisher.

In fact, by some broad measures, investor concern appears to be ahead of formal regulation. When the Carbon Disclosure Project began in 2002, 35 institutional investors endorsed a letter requesting disclosure of GHG emissions through a questionnaire that was distributed to *Fortune* 500 companies.[81] In 2003, 95 institutional investors with $10 trillion in assets endorsed the letter. By 2006, that number reached 211 institutional investors with $31 trillion in assets.[82]

The Sarbanes-Oxley Act of 2002 is also prompting more companies to discuss climate change and associated risks in their annual reports. A critical question, about which there remains some uncertainty, is whether climate concerns are "material" under Security and Exchange Commission (SEC) rules. Some point out that the answer to this question is likely to vary by industry and depends on whether GHG controls are legislated. As one study suggests, "While climate change risks and opportunities are unlikely to have material effects over the short-term…the certifications required by Sarbanes-Oxley will put ongoing pressure on management to account for and disclose, in financial statements or otherwise, any aspect of climate change risk which could be fairly said to be quantifiable."[83] In any case, while only 26 percent of survey respondents believe that GHG emissions are "material" under Sarbanes-Oxley, the vast majority (93 percent) nevertheless consider climate-related risks when making general investment decisions. As Bob Page, Vice President of Sustainable Development at TransAlta, puts it, "Shareholders must understand actions taken to manage GHG and climate risks."

Shareholder understanding and concern about climate issues appears to be growing and the investment community is increasingly concerned about the uncertainty of federal climate policy in the United States. Coal-fired power plants, for example, have projected lifespans of 40 years or more and will almost certainly be affected by future climate regulation, particularly as it relates to feedstock and energy prices. In the current policy void, however, many investors feel they are unable to adequately analyze these potential risks and opportunities. Travis Engen, former CEO of Alcan says, "Some asset-intensive industries are making investments now that have a 30 to 50 year horizon. As CEO, I wanted to make damn sure my investments were good for the future, not just today."[84]

The Mainstream Financial Community is Taking Notice

Mark Tercek, Managing Director, Goldman Sachs

Goldman Sachs takes seriously its responsibility for environmental stewardship and believes that, as a leading global financial institution, it can and should play a constructive role in helping to address the challenges facing the environment by doing what it does best—applying its people, capital and ideas to find effective market-based solutions to critical environmental issues. To guide its efforts, the firm introduced an environmental policy framework in November 2005.[87] The policy framework includes a number of specific elements: using the firm's expertise as a market maker to create more efficient markets for environmental products and services and acting as a liquidity provider by actively trading in those markets; applying research capabilities to examine the impact of environmental risk and the associated business opportunities; developing a deeper understanding of the impact of environmental issues on clients in order to more effectively advise and partner with them; and investing in renewable and alternative energy sources.

Financial experts are quick to differentiate uncertainty from risk. In general terms, "risk is a mathematical distribution of potential outcomes around known parameters, even if the actual parameters and shape of the distribution is in dispute. Uncertainty, on the other hand, involves a lack of information for determining the parameters with which to assess investment risk."[85] As a result, "in Europe, climate regulatory risk can be analyzed because the parameters around policy are generally known. However, in the U.S. not only is the likely future structure of policy not known, but competitive responses by companies to these polices are difficult to estimate."[86]

Though climate risk has yet to play a significant role in valuation, it has begun to play a role in project finance. In the past few years, some of the world's largest investors—including JP Morgan Chase, Goldman Sachs, Citibank and Bank of America—have announced plans to begin including GHG emissions when analyzing potential projects.

More recently, the finance community has begun to seek out potential opportunities created by climate regulation. JP Morgan has committed to creating financing solutions for investments in GHG reductions and low-carbon technologies and Citigroup has released a report on companies whose profits could grow under climate regulation. According to Citigroup, these companies are "the modern equivalent of the companies that sold products to California's panners in the 1849 Gold Rush; the companies providing the means—picks and shovels—to achieve an end, reducing energy usage and cutting GHG emissions in the end."[88] The Citigroup report includes mainstream blue chips like General Electric (which produces components for IGCC systems, wind turbines, and efficient appliances) and more specialized companies such as Itron (which provides energy metering and management technologies) and Kinder Morgan (which develops carbon sequestration technology).

To systematically link climate-related risks and opportunities, Innovest has created the "carbon beta" which uses proprietary data to incorporate three broad factors in corporate finance decisions: (1) the cost of a company's carbon exposure as a percentage of revenues (which can be up to 10 percent at $20 per ton); (2) the company's geographic risk exposure (that is, whether a company operates in a country that has ratified the Kyoto Protocol or in a country, like the United States, where it might be exposed to litigation); and (3) company-specific factors such as energy intensity and technological trajectory (for example, if the company has developed a "silver bullet" technology). As Innovest analyst Doug Morrow has remarked, "Upside and downside exposure to climate change is not yet priced into the fixed-income or equity markets, so there is out-performance potential in a product that uses specialized research to address these factors."[89] In May 2006, Innovest partnered with UBS to offer a bond fund based on its carbon beta methodology.

C. External Resistance

Not all external stakeholders support corporate action on climate; indeed 43 percent of survey respondents encountered external resistance. Of this group, 82 percent cite regulators as a barrier, with some pointing to the lack of clear climate policy as the form of that obstacle. Similarly, according to the consulting firm Deloitte, some executives in the power and utility sector say "the lack of specific policy guidance makes voluntary remedies a guessing game."[90] All survey respondents report efforts to overcome external resistance by lobbying at the national level and 88 percent also lobby at the state level.

Trade associations can be a tool in lobbying efforts (they are used by 62 percent of companies), but many also oppose action on climate change. More than one-third of survey respondents are members of trade associations or other organizations that oppose climate change regulation and 36 percent list trade associations as obstacles to climate action. Instead of discontinuing their membership, however, most companies prefer to work within their trade associations, citing opportunities to inform and influence others as well as to understand other positions on the issue. According to DuPont's DeRuyter, "You should not become overly aggressive if you cannot get agreement. And with the [American Chemistry Council] it can be very

Improving Industry Specifications to Reduce GHG Emissions

Tom Chizmadia, Vice President, Communications and Public Affairs, Holcim (U.S.) Inc.

U.S. cement producers can reduce CO_2 emissions per ton of cement manufactured through the addition of mineral components such as fly ash or slag. However, reluctance to use these cements in construction material specifications and their use in project designs are significant barriers to achieving the reductions. The two organizations whose material specification standards have the most significant impact on the cement industry are the American Society of Testing and Materials (ASTM), and the American Association of State Highway Transportation Officials (AASHTO). Unfortunately, the specification standards that can yield CO_2 reductions in cement are infrequently applied by specification writers, and there has been a lack of consensus between the organizations about the addition of small amounts of limestone to portland cement. ASTM began allowing up to 5 percent limestone in 2004 while the AASHTO standard did not. However, a joint ASTM/AASHTO task group is working toward adopting one national standard for limestone additions.[91] There is little evidence to suggest that specification writers will use a blended or slag cement based solely on its potential to reduce CO_2 emissions. But Holcim (U.S.) Inc., one of the nation's largest manufacturers and suppliers of cement, has been working for over twenty years to demonstrate that a broader application of performance specifications can yield both excellent concrete and lower CO_2 emissions per ton of cement manufactured. Holcim has worked individually and through its trade associations to educate specification agencies, public officials, and customers about the technical merits of cements made with mineral components and to broaden their acceptance in the marketplace. The company has worked with several state departments of transportation about increasing the use of standards that allow composite and blended products, and has provided technical and empirical data to university engineering departments about the use of these materials. The company also is actively engaged in public policy discussions about CO_2 reduction strategies for the cement industry.

hard to get agreement with companies on either end of the spectrum." DuPont takes a cooperative approach, focusing its efforts within organizations that are actively engaging the climate issue, like the Pew Center, the International Climate Change Partnership (ICCP), and the Business Roundtable. Cinergy CEO Rogers has announced that he will adopt a less aggressive stance on climate when he becomes chairman of the Edison Electric Institute (EEI) and will advocate for voluntary rather than mandatory programs when he is speaking for EEI.[92]

Of the case study companies in this book, Whirlpool stands out as the only one that took a more confrontational approach by deciding to withdraw from the American Home Appliance Manufacturers (AHAM) over a difference of opinion on energy-efficiency standards. Whirlpool later rejoined AHAM after changes were made in the organization's bylaws.

Inviting Customers to Take Action on Climate Change

Steven Kline, Vice President, Corporate Environmental and Federal Affairs, PG&E Corporation

PG&E provides electric and natural gas service to more than 15 million people throughout northern and central California and has some of the most environmentally conscious customers and policies in the nation. For example, California's commitment to energy efficiency has allowed per capita energy consumption in the state to remain flat for nearly 30 years, while the state's economy has grown dramatically. The state also has an aggressive renewable portfolio standard for investor-owned utilities (IOUs) and requires IOUs to apply a "greenhouse gas" adder when evaluating bids from power suppliers. As a result of these policies and other company actions, PG&E's carbon emission rate per kwh of generation is among the lowest of any IOU in the country, while its emission rate per kwh of delivered electricity is 60 percent below the national average. When PG&E looked to be responsive to its customers' desires to address climate change and to policymakers' calls for innovative approaches to do more, the company found it needed to look beyond traditional "green" tariff, energy efficiency, and other pricing programs.

In January 2006, PG&E voluntarily proposed a Climate Protection Program through which customers can choose to sign up and pay a small premium on their monthly utility bill to fund independent environmental projects aimed at removing CO_2 from the atmosphere. The first projects will focus on forest restoration and conservation, and the carbon sequestration and emission reductions from those projects will be verified by the California Climate Action Registry (CCAR). Once verified, the reductions will be permanently retired. PG&E expects to enroll approximately 4–5 percent of eligible customers into the program by the end of its third year, and achieve carbon reductions equivalent to taking 350,000 cars off the road for a year. As stated by Thomas Bottorf, PG&E's senior vice president of regulatory relations, "Through this first-of-its-kind demonstration project, we look forward to giving our customers the opportunity to help remove greenhouse gases from the atmosphere while also improving California forests and habitat."

D. Supply-Chain Partnerships

Targeting the supply chain—from customers to suppliers—provides the best opportunity for using outreach to connect climate concerns with a firm's business objectives. Exelon, for example, called on its suppliers to reduce emissions when it adopted a voluntary GHG reduction goal. All the suppliers who responded to Exelon's initial outreach reported that they were already engaged in some related activity, even if they had not originally characterized their efforts as part of a climate strategy. Examples included recycling or efficiency, renewable energy investments, transportation programs, and marketing programs.

Interface is working with the United Parcel Service to better understand GHG impacts associated with parcel freight transportation. Interface has also relied heavily on Invista, a large supplier of fiber for the

Industry-Wide Action to Reduce GHG Emissions

Tim Higgs, Environmental Engineer, Corporate Environmental Department, Intel

In the early 1990's, perfluorinated compounds (PFCs) in the semiconductor industry became an issue of concern due to their high global warming potential, which is thousands of times that of CO_2 (GWPs for PFCs range from 5,000 to 25,000). While there was no legal or regulatory requirement to reduce emissions, climate change was then beginning to emerge as a significant environmental concern and the industry recognized its obligation to demonstrate environmentally responsible management of these materials. So, Intel and the rest of the semiconductor industry developed a worldwide agreement to publicly commit to reducing PFC emissions 10 percent below 1995 levels by 2010. Intel took a leadership role on this initiative because its own environmental health and safety policy states that it will implement programs that go beyond regulatory requirements where appropriate. This was one of those times. Many were concerned that new regulations or even material bans were possible if the industry did not demonstrate leadership on the issue. In fact, several key PFC suppliers had publicly stated that they would require users to demonstrate proper management of PFCs before they would agree to sell to them. This strategy has proven successful. Bans or restrictions on high-GWP fluorocarbons have been proposed in recent years, primarily in the E.U. But these proposals have not focused on the semiconductor industry in large part because of this PFC agreement; the industry is viewed as taking its responsibilities on climate change seriously and acting to reduce emissions on its own. At this writing, Intel's own manufacturing has increased more than two and a half times since 1995 while its PFC emissions have remained roughly equal to 1995 levels, putting the company well on track to meeting its target by 2010.

company's carpet business, to develop its climate-neutral Cool Carpet™ and Cool Fuel™ programs. In 2003 and 2004 Invista provided over 62,000 metric tons of certified CO_2e reduction credits from its Orange, Texas plant. Going further, Wal-Mart recently announced that it will begin to require emission reductions from its suppliers.

Companies often work closely with business partners on climate-related activities. For example, Whirlpool worked with retailers (like Lowes and Sears) and with consumers to address misconceptions about the efficacy of energy-efficient appliances and to educate people about their benefits, including their average five-year payback period. Whirlpool also worked with Proctor & Gamble to ensure that detergents suitable for their more efficient machines were available and to educate consumers on their use. Finally, the company was pivotal in convincing *Consumer Reports* magazine to include energy efficiency in its appliance rankings.

Some firms participate in multi-company consortia to advance their climate objectives. For example, DuPont leads the Integrated Corn Bio Refinery consortium, which includes private, public, and academic participants and has been awarded $19 million in matching funds from the U.S. DOE. Similarly, Alcoa works with the Curbside

Industry-Driven Innovation on Climate Change

William Sisson, Director, Sustainability, WBCSD Buildings Program, United Technologies Corporation

Buildings account for 40 percent of global energy demand and nearly 37 percent of total CO_2 emissions. United Technologies Corp. and Lafarge are in the early stages of partnering with other companies through the World Business Council for Sustainable Development to lead a global industry project that will provide a response, framework, and timeline to advance effective and responsible building technologies for a carbon-constrained world. The goal is to influence a market transformation by 2050 across all business industry segments. "The biggest and shortest-term impact on greenhouse gas emissions, which many hold to be the greatest sustainability problem we face, is reducing energy consumption by marked improvements in installed product efficiencies," says UTC's Chief Executive Officer George David. "The best sustainability efforts, like everything else in human endeavor, are those coming from marketplaces and not mandates." To spur industry-wide investment in climate change technologies, governments must commit significant financial incentives and R&D. Some of these technologies already are under development by UTC, including collaborative information tools that facilitate energy-efficient and economically viable buildings; technologies that increase heating/cooling system performance and efficiency; information infrastructures that better manage fire and security systems; elevator regenerable power drives; and renewable and fuel cell technologies for on-site power co-generation. With stronger federal support for such R&D activities, UTC believes the technologies needed for a self-sufficient, energy-efficient building are right around the corner.

Value Partnership (CVP) to educate the public and promote recycling through existing curbside collection channels. Finally, to demonstrate the value of terrestrial carbon sequestration, AEP, Cinergy, Entergy, Exelon, Wisconsin Energy, and other power companies are members of PowerTree Carbon Company, LLC, which plants trees in critical habitats in the Lower Mississippi River Valley.

Concerted efforts to address climate change allow companies to stand out as industry leaders. Alcoa and DuPont were cited in *Business Week*[93] for their climate-change accomplishments, and Ceres scored many BELC companies[94] highly in one of its recent reports.[95] Interface has a long received external recognition for its environmental work, including from the *Progressive Investor, Business Ethics*, *GlobeScan*, and *Global Finance,* among others. Recognition for climate leadership, in turn, can create further business opportunities. Alcoa executives, for example, were approached by Toyota for possible business ventures after the two companies (along with BP) were singled out by Innovest as the world's top three most sustainable companies.

Conclusions

The prospect of GHG controls is already altering existing markets and creating new ones.[96] As in any market transition there are risks and opportunities and there will be winners and losers. All companies will be affected to varying degrees, and all have a managerial and fiduciary obligation to at least assess their business exposure to decide whether climate-related action is prudent.[97] The companies in this book believe a proactive approach is necessary to prepare for the coming market transformation and that doing nothing means missing myriad near-term financial opportunities and setting themselves up for long-term political, operational, and financial challenges. Looking ahead, these companies identify three key drivers that will hasten the transformation to a carbon-constrained world.

The first driver is very clearly the **establishment of regulations**. When policy is set, the business landscape will change. Market signals will emerge that will drive technology and products toward a reduced carbon footprint. Companies hope to be fully prepared for that transformation and, ideally, to have a hand in shaping the policy.

The second driver is **rising energy prices** which will have different implications for different industries and companies. Rising energy prices help companies like Whirlpool or Intel promote more energy-efficient products in the marketplace. Conversely, they pose a threat to energy-intensive industries such as aluminum and cement. According to Cinergy's Leahy, "The sudden ramp up in energy prices may be changing the political landscape around this issue. On the one hand, it makes it easier to talk conservation but harder to talk about using a carbon price to pull new technologies along. People haven't made the connection between the fact that energy prices move up and down all the time—sometimes a lot—and the fact that an entry level carbon price shouldn't be that noticeable to consumers, yet it will change behavior at the margin."

"I worry that we are using 100 year-old technology. There will be a transformative technology. At what point will our generation and transmission lines become obsolete?"

The third driver companies are watching is growing **interest within the investment community**. Baxter's Meissen sees "an increased volume of requests from investors for companies to disclose GHG data, define climate strategies, and report progress in reducing emissions."

In sum, climate considerations are already altering the business environment in ways that are real and yet still fluid. The rules of the game are changing and companies ignore these changes at their peril. For example, Cinergy CEO Rogers says, "I worry that we are using 100 year-old technology. There will be a transformative technology. At what point will our generation and transmission lines become obsolete? There are a lot of things you might do, if you think there will be a new technology in 25 years. You need to hit your numbers with a short-term view, but you need to run your company with a long-term view." Shell's Hone has similar thoughts. "The key is both influencing the rules of the game and timing your transformation to a new carbon-constrained strategy. It's knowing what the next technology for energy production is, and transforming when the market is ready to reward it. We're not going to get out of the oil business in the near term." But, Hone says, you have to ask, "What is the iPod® for energy? Is it out there? You have to be on watch."

Case Studies

Case Studies

Managing "Stroke of the Pen" Risk

Cinergy*

Cinergy's heavy reliance upon coal combustion for electricity generation makes it particularly vulnerable to carbon regulation. Yet, according to Chairman and CEO Jim Rogers, addressing greenhouse gas (GHG) emissions is not only the ethically right thing to do; it is also a smart business decision. Rogers believes that U.S. industry will soon face domestic carbon constraints, a prediction that presents Cinergy with a serious strategic challenge. While climate change is a long-term problem, many industries need short-term regulatory and market clarity in order to properly value potential investments. For companies like Cinergy within the power sector, the future of climate policy and carbon regulation will affect strategic decision-making about investments in new generating capacity that have an expected life of 40 or 50 years.

Table 6

Cinergy's Footprint (2005)

Headquarters:	Cincinnati, OH
Revenues:	$4.6 billion
Employees:	7,842
Percentage of Emissions In Kyoto-Ratified Countries:	0 percent
Direct CO_2e Emissions Legacy Generating Units:	58.2 MMtons*
Cinergy Solutions Projects:	2.6 MMtons
Other Direct CO_2e Emissions:	0.3 MMtons
Aggregate CO_2e Emissions**:	61.1 MMtons
Target:	5 percent reduction in GHG below 2000 levels by 2010-2012
Year Target Set:	2003

* Million metric tons.

**Cinergy does not track indirect emissions resulting from power purchases nor does it calculate emissions from product use.

"The greatest risk we face is '*stroke of the pen*' risk, the risk that a regulator or congressman signing a law can change the value of our assets overnight," says Rogers. "If there is a high probability that there will be regulation, you try to position yourself to influence the outcome." Cinergy is actively managing this regulatory risk through its voluntary GHG emission reduction program and its aggressive leadership role within the utility industry. These actions make the company a legitimate participant in the national policy debate, creating the opportunity to work with government, trade associations, environmental organizations and other stakeholder groups to help shape legislation on GHG emissions. But while Rogers leads Cinergy with a long-term focus, he does not feel that the company can take definitive action on climate change until there are both clear regulatory and market signals to do so. As Kevin Leahy, Managing Director, Climate Policy, explains, "The technologies will emerge when CO_2 has a price signal. All we need is a market signal to act, and that market signal will be created by regulation."

* We would like to thank Eric Kuhn, Kevin Leahy, David Maltz, Darlene Radcliffe, Jim Rogers, Catherine Stempien, and John Stowell for their contributions to this case study.

Company Profile

Cinergy is one of the leading diversified energy companies in the United States, with 2004 revenues exceeding \$4.6 billion and a workforce of 7,842 employees. The company was created in 1994 through the merger of Cincinnati Gas & Electric (CG&E) and PSI Energy, Inc., the largest electric utility in Indiana. Cinergy is currently organized into two core businesses: Regulated Operations and Commercial Businesses.

The Regulated Operations unit consists of PSI's regulated generation, transmission and distribution operations, and CG&E's regulated electric and gas transmission and distribution systems. This unit plans, constructs, operates and maintains Cinergy's transmission and distribution systems, and delivers gas and electric energy to consumers. It owns over 7,000 megawatts (MW) of electric generating capacity serving 1.5 million electric customers, and operates 9,200 miles of gas mains and service lines that serve about 500,000 customers.[98]

The Commercial Businesses unit is comprised of the wholesale generation and energy marketing/trading operations. This includes CG&E's 6,300 MW of electric generating capacity in Ohio, which was deregulated in 2001. The wholesale generation division also includes the subsidiary company Cinergy Solutions (Solutions), which owns or operates 27 cogeneration projects with over 5,400 MW of electric generating capacity and performs energy risk management analyses, provides customized energy solutions and is responsible for all international operations.[99] Solutions' projects usually entail taking an ownership position in the energy production or distribution facilities of strategic partners and reworking the facility to improve energy efficiency and environmental performance. In addition to producing bottom-line revenues, these projects usually generate GHG reduction benefits as well.

In 2004 Cinergy generated 69 million megawatt hours of electricity, 98 percent of which were generated from the combustion of 28.2 million tons of coal, approximately 2.8 percent of the total 1.016 billion tons of coal consumed for electric power in the United States.[100] Cinergy's 2004 CO_2 equivalent (CO_2e) emissions totaled 68.6 million metric tons, representing almost one percent of total CO_2e emissions in the U.S.[101] The majority of these emissions (94 percent) are from "legacy generating units," those electric generating plants that were part of the original CG&E and PSI utility systems, as well as those electric generating plants acquired by the unregulated merchant group that are not Solutions projects. These figures have changed, as Cinergy has been acquired by Duke Energy through a \$9 billion stock swap (see "Cinergy's Merger with Duke Energy" on page 77).

Climate Change Program Implementation

Cinergy began paying attention to climate change with a study in the early 1990's by ICF Consulting on the feasibility of adopting an internal CO_2 cap. Given the coincident activities surrounding the CG&E/PSI merger, the study only served to awaken concern within the company. GHG goal development was initiated in 1993 with Cinergy's participation in the Edison Electric Institute/U.S. Department of Energy (DOE) Climate Challenge. In September 2003, Cinergy formally announced its voluntary GHG emissions reduction program, with the goal of reducing annual emissions to five percent below the 2000 baseline for the years 2010 through 2012. The company's decision to more aggressively

Cinergy's Signposts

Signpost #1: States are taking action.

Signpost #2: An increasing number of U.S. Senators are expressing concern about global warming.

Signpost #3: The Kyoto Protocol was ratified and became law on February 16, 2005.

Signpost #4: A growing number of shareholder groups are asking companies to quantify the risks associated with GHG emissions.

Signpost #5: CO_2 and GHG emissions trading markets are developing in Europe and the United States.

Signpost #6: Global warming is becoming part of our everyday consciousness.

embrace climate change was made possible by three forces converging: an internal management push, pull from external stakeholders and technological developments that would allow the company to move forward in a carbon-constrained world.

Internal Management Push. Chairman and CEO Jim Rogers leads Cinergy with a long-term view and an approach that is rooted in stewardship. Given the expected 40 to 50 year lifespan of investments in generating capacity and the regulated nature of the industry, long-term planning is common for utilities. However, the principles of stewardship employed by Rogers are rare. "When your time horizon is short, you're thinking 'stonewall it and it won't happen on your watch,'" Says Rogers. "If you are a steward, you make decisions on a longer time horizon, looking beyond your own tenure. When you think of it that way, your view changes. We look 20, 30, 50 years down the road."

Today, when Rogers looks out over the business horizon, he sees six "signposts" indicating that climate change is an issue to be dealt with head on (see "Cinergy's Signposts" on this page[102]). Notably absent from this list is scientific research and analysis. According to Rogers, "Our decisions are purely business based. The science is interesting, but not truly relevant for our purposes." Based upon these trends, he believes it is his responsibility to prepare the company for the likelihood of operating in a carbon-constrained world.

Cinergy deals with climate change as a long-term systematic effort primarily through capital investments and a focused public policy stance. This approach is well suited to the utility industry and aligned with the long-term nature of the climate change issue. Because climate change is caused by the concentration of long-lived GHGs in the atmosphere, there is reason to begin action but not immediate draconian reductions. The mantra is "slow, stop and reverse the growth of emissions." Yet, according to Eric Kuhn, Principle Environmental Scientist, "There is a real commitment on Jim Rogers' part to provide resources for this issue. CEO buy-in is critical, especially for a voluntary program."

Rogers' leadership style infuses the corporation with a strong focus on stakeholder engagement and transparency. His varied background and credentials lend legitimacy to his messages and engender trust from his audiences. Prior to joining PSI in 1988, he acted as an intervener on behalf of consumers in gas, electric and telephone rate cases in the Commonwealth of Kentucky, served as Deputy General Counsel for Litigation and Enforcement of the Federal Energy Regulatory Commission (FERC), and legally represented energy companies before the FERC, the Department of Energy, various Congressional committees and federal courts. Rogers has testified before Congressional Committees 13 times since 1989, on issues ranging from the environment to national energy strategy to industry restructuring.

The culture of stakeholder engagement dates back to when Rogers became head of Public Service Indiana (PSI) in 1988. At that time the company had a failed nuclear program, very poor relations with customers and was nearly bankrupt. Rogers introduced a strategy to improve relations through meaningful engagement with environmentalists, consumers and industrial groups in the state. Having a dialogue and listening with an open mind has developed trust from stakeholders, which has proven to be an asset for the company in efforts ranging from rate cases to locating infrastructure development. This credibility has extended into the policy arena, allowing Cinergy to base discussion on climate change on what it views as an economically rational foundation. Cinergy believes its collaborative approach is good for all of its stakeholders, including investors, customers, employees, policymakers, regulators, suppliers, partners and communities.

In fact, stakeholder engagement played a significant role in stimulating a more public position from the company on climate change. Early collaboration with the U.S. DOE on the Climate Challenge program and on-going interaction with policy makers on three air pollutant issues (sulfur dioxide, nitrous oxides and mercury) provided insight into the future of carbon regulation. Subsequent to these efforts, Cinergy made a commitment to participate in the U.S. EPA's Climate Leaders Program.

Pull from external stakeholders. In 2002, the Committee on Mission Responsibility through Investment (MRTI) of the Presbyterian Church (USA) submitted a shareholder resolution requesting that Cinergy provide information on GHG emissions and disclose the risks associated with climate change. Cinergy appealed to the Securities and Exchange Commission and was granted no-action relief. After MRTI tried again in early 2003, the company chose to reach out and engage in discussions that ultimately led to MRTI withdrawing the proposal. This dialogue also resulted in the development of a plan to disclose Cinergy's risks related to climate regulation.

In September, 2003, the company formally announced its internal GHG reduction program, a response to both the Climate Leaders Program commitment and the intervention by MRTI. In February, 2004, the company announced it would partner with MRTI to develop the *Air Issues Report to Stakeholders* (*AIRS*). The December 2004 issuance of *AIRS* was a watershed moment for Cinergy. The report provided a broader analysis of the company's risks related to climate change and other emissions, with a thorough discussion of the linkage between energy, economics and the environment. The effort also represented a more public positioning on climate change and a culmination of analysis that had begun years earlier.

Technological developments. Heavy reliance on coal exposes Cinergy to regulatory risk in any form of carbon regime. Despite this fact, coal's abundance and low cost in the United States leads the company to believe that coal will continue to be central to the country's longer-term fuel mix. Cinergy's work with environmentalists gave it an early indication of a potential to break the carbon-environmental impasse; some environmentalists were warming to the idea of coal being part of the solution.

The most promising means currently available for utilizing coal in a carbon-constrained world is through the implementation of Integrated Gasification Combined Cycle (IGCC) technology combined with Carbon Capture and

Sequestration (CCS). The coal gasification process converts coal into a synthesis gas (syngas) and produces steam. The hot syngas is processed to remove sulfur compounds, mercury and particulate matter before it is used to fuel a combustion turbine generator. The heat in the exhaust gases from the combustion turbine is recovered to generate additional steam. This steam, along with that from the syngas process, then drives a steam turbine generator to produce electricity. The technology has the potential to capture CO_2 much more economically than other coal technologies because a concentrated stream of CO_2 can be more readily removed from the syngas of an IGCC plant. Captured CO_2 would then be injected deep underground for geologic sequestration. Industry analysts estimate that carbon capture could add as much as 72 percent to the cost of electricity from a conventional pulverized coal plant, 60 percent to the cost of a natural gas combined cycle plant, but only 25 percent to the cost of electricity from an IGCC plant.[103]

The company has been involved in IGCC since the early 1990's when it built one of the first demonstration plants in the United States in partnership with the U.S. DOE through the Clean Coal Technology Demonstration Program. The West Terre Haute, Indiana plant is still in operation today with Cinergy purchasing syngas from it for one of the units at its Wabash River Station. In 2004, Cinergy entered into an agreement with GE Energy and Bechtel Corporation to study the feasibility of a commercial-scale (600 MW) IGCC generating station. Although various sites were evaluated as potential candidates, Cinergy's preferred IGCC site is the current location of a 160 MW pulverized coal plant near Edwardsport, Indiana built in the late 1940's. Given the importance of the climate change issue and the ability to continue to use coal, geologic sequestration potential was included as one of the siting criteria for the first time as part of the company's internal evaluation. A Front End Engineering and Design (FEED) study is being undertaken and should provide enough detailed design and cost information for a decision to be made whether or not to move ahead with the plant by late 2006.

Ultimately, Cinergy believes that resolving the climate change issue will require a paradigm shift regarding the technologies employed to refine and use energy. The types of technologies being discussed today and deployed over the next 20 to 30 years will all continue to utilize fossil fuel as their source of energy; even hydrogen would likely come from fossil fuels. Although they are more energy efficient and have the capability to capture CO_2, they are only stopgap or bridging technologies to be used until low- or zero-carbon technologies are developed and deployed in the second half of this century.

But, notes Kuhn, "We are not a technology developer or owner. We are a customer for new technologies to enable us to economically operate our plants and/or produce electricity. We will however work with partners to provide test sites and assistance. But we'll likely not be the owner of resulting patents. We know intuitively that the cost of reductions could be huge so that the pennies that we are investing in research today could have tremendous returns in the future if only a small portion of the costs are reduced."

Climate Program. Cinergy's GHG Management Goal of five percent below 2000 levels for the period 2010 through 2012 was developed to position the company to take meaningful actions on GHG emissions and provide the company with credibility to lead the climate change policy debate. But developing the goal first involved a risk assessment process, performed by Cinergy's risk management and portfolio optimization teams, which examined a variety of options for action.

Once an optimal goal was selected, it was reviewed by various non-governmental organizations (NGOs) and with that input, revised goals were presented to Cinergy's senior management. Many were unsure of the wisdom of setting such a goal, but most were persuaded that the strategic positioning and organizational learning were worth the associated risks. The goals were presented to Cinergy's Board of Directors as a matter of course, although not for official adoption. Similar to DuPont's response to both CFCs and GHGs, Cinergy set a target that was a stretch, not knowing precisely how it would achieve it.

The first step in implementing the new goal was performing an assessment of the baseline year-2000 GHG emissions. This effort was completed in 2004 and reviewed by Environmental Defense, who acted as an independent third party to add validity to the process. Environmental Defense has reviewed Cinergy's definition of its corporate emissions footprint, approved how GHG reductions are identified and measured, evaluated the company's implementation of the GHG fund, and serves as an ex-officio member of the GHG Management Committee that is charged with implementation of Cinergy's GHG goal. Cinergy has not yet engaged a third party auditor to verify its calculations, but plans to do so in 2006. Baseline year-2000 emissions were calculated to be 73.8 million metric tons CO_2e (see Table 7).[104]

Given historical trends in energy demand, Cinergy's GHG Management Goal of a five percent reduction translates to an emissions level of approximately 70 million metric tons per year.[105] The goal was reviewed by EPA Climate Leaders staff, who determined, based on their own projections for electricity demand in the region, that the proposed goal was substantial. During the three year period 2010 through 2012, approximately 30 million metric tons of CO_2e emissions reductions would be achieved.[106]

Table 7

Baseline year 2000 CO_2 Equivalent Emissions

Source of Emissions	Tons CO_2e	Percent of Total
Legacy Electricity Generating Units	69,768,000	94.48
Fugitive Natural Gas	409,000	0.55
Cinergy Solutions Projects	3,454,000	4.68
Fleet Vehicles	36,000	0.05
SF_6 Emissions	176,000	0.24
Total	**73,843,000**	**100.00**

Reductions will come from the company's regulated and non-regulated electricity generating units, combined heat and power (CHP) facilities, natural gas distribution system, vehicle fleet operations and other operations that emit significant amounts of GHGs. Cinergy takes credit for emission reductions from its Solutions business, but only if it has an ownership position and operates the facility. The emission credits are not prorated based on a percentage of ownership since Cinergy is taking responsibility for all of the GHG emissions from the facility. Cinergy operates, but does not own, a number of industrial power generation and CHP facilities. When Cinergy has no control over capital investments or operational changes at these units, their emissions are not included in the GHG baseline. Unless ownership passes to Cinergy, such emissions will not be included in future measures. Furthermore, Cinergy does not track the indirect emissions that result from power purchases, as it is difficult to determine the origin of electricity purchased by traders. Finally, emissions from the mining and transport of coal are not included in the calculations.

Cinergy intends to achieve at least two-thirds of emission reductions "on-system" (or within its operations), and up to one-third "off-system."[107] On-system emission reductions involve projects that impact Cinergy's direct emissions. Examples include: CO_2 emissions from smoke stacks and vehicular tailpipe CO_2 emissions, methane emissions from the natural gas distribution system, or SF_6 emissions from the transmission and distribution system. Examples of off-system reductions include forestry projects and research and development projects. Implementing both on-system and off-system projects will generate experience and knowledge regarding in-house technical capabilities for reducing GHG emissions as well as real-time data regarding the cost-effectiveness of such efforts. By taking these actions now, Cinergy will be better prepared to contribute to the policy discussion and to operate in a carbon-constrained future.

As emissions reductions are achieved, they are reported to the U.S. DOE's Energy Information Administration (EIA) through the 1605(b) reporting system and to the U.S. Environmental Protection Agency (EPA) as part of Cinergy's commitment under the Climate Leaders program. Cinergy feels strongly that early actors must receive credit for their voluntary reductions when legislation is ultimately passed.

Carbon dioxide is directly measured at generating units equipped with continuous emissions monitors (CEMs). For stations not equipped with CEMs, estimates are calculated using the BTU value of the fuel consumed multiplied by the pounds of CO_2 emitted per million BTU as provided through the DOE's EIA 1605(b) reporting program.

Measurement and verification of biological CO_2 sequestered by tree plantings undertaken by Cinergy begins with the identification of measurement plots for testing. Within each sample measurement plot, tree volumes, underbrush and soils are measured for carbon content. The measurements are repeated at regular intervals, data is extrapolated between years when the measurement plots are surveyed and the measurement results are applied to the entire acreage of plantings. This process provides a statistical confidence level of 95 percent.

Organizational Integration

In the years 2004 and 2005, Cinergy budgeted $3 million (what Leahy calls "tuition to learn") for projects to reduce GHG emissions, the first two installments of seven comprising the total $21 million GHG fund through the end of the decade. This budget is managed by the GHG Management Committee (the Committee), which is comprised of ten senior representatives from business areas that would be affected by GHG restrictions (legislation) and one ex-officio member, Environmental Defense. Annually, GHG-reducing and offsetting projects are solicited throughout the company and are open to any employee who would like to propose a project. Project proposals are limited to five pages in length and include a description of how the project will reduce GHG emissions, quantification of projected reductions, evaluation of the project's permanence, and an analysis of cost estimates for the project. Another critical factor is whether or not the project would be implemented without GHG Funds. Projects are reviewed, evaluated and ranked by staff using criteria established by the Committee. The projects are then presented to the Committee for their consideration and funding.

In 2004 and 2005, the Committee received over 150 project proposals. The majority of on-system projects were small efficiency projects in the power plants. Other on-system projects included wind and solar demonstration projects, the purchase of four hybrid vehicles for the Cinergy transportation fleet, and customer end use electric efficiency projects. Customer electric efficiency projects are considered on-system because they reduce the CO_2 emissions from Cinergy's power plants. Examples of off-system projects included tree planting and the funding of research and development projects in the areas of carbon sequestration, biomass fuels, and renewable energy generation.

In evaluating potential projects, Cinergy does not use a shadow price for carbon, largely because internal sentiment is that regulation is too remote and uncertain to reliably quantify a price. Another reason not to use a particular cutoff price for carbon is the secondary benefits commonly associated with the efficiency projects, such as reduced fuel consumption and reduced SO_2 and NO_x emissions. Preliminary data collected for the power plant efficiency projects implemented in 2004 indicate that the projects actually return value to the company in the form of fuel savings and generation of SO_2 and NO_x allowances. These projects were considered "low-hanging fruit" but as the company moves forward with its climate change program, reductions are expected to become more costly.

The criteria currently used to evaluate project proposals are more subjective than objective, including considerations such as the age of the facility and its availability rate. Ultimately, the Committee is interested in the cost per ton of CO_2e emissions reduced, but it also considers issues such as project replicability, longevity of reductions achieved, and whether funding sources other than those related to GHG would be available. However, the cost data being gathered is part of the institutional learning desired by the committee, generating hard data from historical actions on available reductions at various price levels. This has value internally as well as in policy debates.

According to Kuhn, many of the on-system reductions have been projects "that were on the cutting room floor because they did not meet internal rate of return criteria." These projects had been previously forgone because the return on modest efficiency gains, in the form of fuel cost savings, was negligible given low coal prices. "However," says Kuhn, "these projects become attractive when the value of GHG emission reductions is taken into account."

Of the \$6 million allocated in 2004 and 2005, \$4.4 million (73 percent) was invested in on-system projects and \$1.6 million (27 percent) funded off-system projects, reducing annual CO_2e emissions by approximately 600,000 and 25,000 metric tons respectively. While it is not fully accurate to calculate a cost per ton from these figures due to the research and development projects that are included, Cinergy estimates that the actual average cost per ton of CO_2e emission reductions was \$8.28 in 2004 (on-system reductions averaged \$6.43 and off-system reductions

Table 8

Cinergy's 2004 **GHG Fund** Projects

Project	Total Incremental Funds	Annual Tons of CO_2 Reduced	Average \$/ton CO_2 (2004-2009 projected)
On-System			
Heat Rate Improvement Projects at Generation Stations	$1,940,000	349,882	$1.11
Markland Dam Software Upgrade	$285,000	7,400	$7.70
Hybrid Cars	$20,000	26	$153.85
Renewable Energy Demonstration Projects *	$55,000	35	$314.29
Off-System			
The Nature Conservancy Reforestation Project	$180,000	1,000	$36.00
Vestar-Oldenburg Academy Energy Conservation Project *	$90,000	62	$290.32
Cincinnati Zoo Education Center Solar Project *	$150,000	33	$909.09
EPRI Research Project	$250,000	---	
Total All Projects	$2,970,000	358,438	$1.66
On-System Projects and Reductions	$2,300,000	77.4 percent	
Off-System Projects and Reductions	$670,000	22.6 percent	

* Small demonstration projects are more expensive than the costs per ton that Cinergy would accept for full scale utility projects.

averaged $59.00) and $12.49 in 2005 (see Table 8). Cinergy has reviewed its reduction calculation methods with Environmental Defense and EPA Climate Leaders staff, and has pledged to hire a third party auditor to verify emissions reductions and provide assurance that figures and estimates are accurate for meeting its period 2010 to 2012 goal.

Looking more long-term, Cinergy is examining the potential of larger scale renewable energy sources in its service area, including wind, solar and biogas/biomass. But, according to Leahy, "Investment options depend in part on what one believes will happen on the technology front when regulation is set. For now, plant efficiency improvements will be first. These will be followed by methane from leaking pipelines and landfills, biomass co-fire in existing coal plants, and upgrades in renewables as possible. Tree planting will be part of the mix, but less than originally assumed as it is more costly than originally thought. There may be technologies like algae-based scrubbers to lower CO_2 from existing plants—though this is very early stage—that will be useful for existing plants."

Some modest funding has been allocated to the development of renewable energy generating capacity, an energy conservation project, and carbon sequestration.[108] However, it is not believed that renewable energy sources will play a significant role in the voluntary GHG emissions reduction program, primarily due to their intermittent characteristics. When renewable energy sources are dispatched in regions where Cinergy operates, economics dictate that the most likely impact is displacement of a gas-fueled unit, rather than a coal-fired unit. However, should GHG legislation be passed, such technologies would become more competitive in a rising wholesale electricity market, and therefore could also become a more viable part of Cinergy's generating portfolio.

That said, not all projects are chosen for low-cost emission reductions or long-term research value. Some are chosen for their symbolic or educational value. For example, the company's purchase of hybrid vehicles for its fleet does not represent the most cost-effective GHG emissions reductions available, but they do succeed in making the program tangible to employees and stimulating conversation.

Overall, the corporate culture of stewardship, the leadership of Jim Rogers, and the structure of the program have all been critical in garnering internal support for the climate change program. Naturally, having capital available to fund projects in a time of capital constraints makes the program much more real for staff working at the plant level. But the most critical component of Cinergy's program implementation, according to John Stowell, VP of Federal Legislative Affairs, Environmental Strategy & Sustainability, has been communication. "Internal and external communications are part of the culture at Cinergy," says Stowell. "Plant managers know about this program. We have meetings with them, and Jim Rogers discusses the issue often."

External Outreach

External communication is an on-going component of Cinergy's GHG reduction program as well. In fact, it is such an integral part of the company's on-going initiatives and strategy already discussed that treating it as a separate initiative is not completely correct. Cinergy actively engages stakeholders to keep them informed and involved throughout the policy discussion and also to gather important feedback. In reality, the company finds the nuts and bolts of the program are of most interest to other specialists, while the wider public is interested in Cinergy's policy position and endorsement of regulation.

One way Cinergy began to engage its many stakeholders on climate change was through a third party consultant who conducted interviews which were published in the 2004 Annual Report titled *Global Warming: Can We Find Common Ground?* Taken as a whole, they led to a number of conclusions that reflect the core of Cinergy's approach to climate change: global warming is a complex problem that must be dealt with holistically; time is of the essence; the customer is still the top priority; good corporate governance is based on stewardship; and uncertainty will likely persist on this issue.[109]

But the challenge the company discovered in reaching out to stakeholders was finding a balance between the short-term interests of some groups of investors focused on quarterly earnings results, and the long-term interests of other groups such as employees, customers and communities. According to Rogers, "It's important to deliver for the investor, but when running your company from a stakeholder perspective, you include customers, communities, everyone. You need to raise rates slowly for the customer. You often need to make decisions that do not necessarily maximize the next quarter." The company has found that, because the financial risk associated with climate change is still uncertain, institutional investors are not as interested in this issue as they are in the prospects for near-term financial results.

Policy Perspectives

The uncertain regulatory environment flows through to uncertainty regarding the value of Cinergy's assets. It also makes it very challenging to evaluate large capital investments going forward. To help resolve this uncertainty, the company has laid out a number of broad criteria that it believes future regulation should encompass. GHG policy should focus on all sectors of the economy, embrace market-based cap-and-trade principles combined with a "safety valve," and be neutral to fuel type. In addition, compliance flexibility, including off-system reductions, is critical to finding a least-cost solution. Finally, GHG policies should be international.

Ultimately, Cinergy believes the policy should take steps to slow, stop and then reduce emissions growth while promoting public-private partnerships for the development of technology solutions (such as IGCC with CCS). The cost for individual companies of complying with GHG regulation will depend upon the timetable for implementation, emissions reduction requirements, allowance allocations, the impact on fuel prices and ultimately the form of the regulation. It is believed that a cap-and-trade program would be less expensive than a command and control approach.[110]

Cinergy communicates this message to lawmakers through the normal channels of the regulatory and legislative processes, including meetings, discussions at conferences and public statements. The company is not alone in its policy stance; utilities such as Exelon, Entergy and PNM have taken similar positions. According to Leahy, "What is important is that lawmakers know that even some coal-fired utilities think it is possible to deal with the climate problem without harming the economy. We've spent more time working on this problem and so have a better understanding of it than most. Our job now is to help other firms by being open with what we've found—facts are friendly." Industry groups such as the Edison Electric Institute provide a forum for CEOs to share perspectives and hear from experts. When Rogers takes the rotating Chairman's position in June, 2006, he hopes to help the organization move toward a broad consensus regarding climate change; perhaps one that focuses on opportunities, not just risks.

Challenges Ahead

Cinergy's strategy was designed to position the company as an industry leader on climate change. A challenge with the May 2005 Cinergy/Duke Energy merger has been to develop a climate position that best suits the combined entity. Fortunately, Paul Anderson, former CEO of Duke Energy, and Jim Rogers, former CEO of Cinergy and now CEO of Duke, both agreed that action needed to be taken on climate change as soon as possible. Although the legacy companies' views differed on the details of the actions needed, the fact that both companies took a leadership stand on climate change policy helped with the development of the policy for the new Duke Energy.

During the first year after the merger, the company formulated and published its position on climate change and joined the U.S. Climate Action Partnership, a group of companies and environmental groups that developed a comprehensive public policy position on climate change and advocate that policy with leaders in Washington. Duke Energy also developed a new company-wide GHG emissions inventory and conducted analyses to determine if Duke Energy should have a voluntary greenhouse gas reduction goal and what form that goal should take. A key learning is that population growth in the Carolinas is fueling the need for additional electric generation in the near-term with a corresponding increase in CO_2 emissions from Duke's generation fleet. So, Duke has set a goal

Cinergy's Merger with Duke Energy

In May 2005, Cinergy and Duke Energy announced they would merge in an all stock transaction. The combined company retains the Duke Energy name, and is headquartered in Charlotte, NC, the home of the much larger Duke Energy (2004 revenues of $22.5 billion and generation capacity of 32,000 MW).[111]

The merger is attractive on many dimensions, climate change being one of them. Rogers feels that the strong cultural fit between the two utilities assures that efforts on climate change will continue. Duke Energy CEO Paul Anderson (who has become Chairman of the combined company while Rogers has taken over as President and CEO) "has already socialized the issue at Duke," says Rogers, "my assignment is to continue to lead on it."

Synergies between the two companies' fuel diversity may help that process along. For example, Duke Energy's 3,600 MW of gas fired capacity located in the Midwest has not been profitable for Duke in the past. But these assets could be utilized immediately by Cinergy to meet system capacity requirements. If gas prices were to drop significantly, they could also reduce carbon emissions by shifting generation away from older coal-fired units, thus creating a partial hedge.

Another important aspect of this merger is nuclear power. Rogers explains, "If you think about a carbon-constrained world and our need for energy, nuclear may be an option for the future." However, both legacy companies that formed Cinergy (PSI and CG&E) had failed attempts at building nuclear capacity. Rogers continues, "Given our history, nuclear was not an option for us; coal and gas were it. Combining with Duke, one of the best nuclear operators in the country, gives us the assets and expertise to work in a future where nuclear is an option."

Despite these benefits, the risks associated with climate change were not part of the asset valuation process. Rogers explains, "They are regulated in rate base, as are we. Intrinsic value does not really change with carbon regulation because the cost would be passed through to rate payers. The [non-regulated] Ohio assets would change in value, but with their very low variable costs, they could remain competitive with a carbon charge." The larger picture shows that the portfolio of the combined company will be more diverse, lowering the regulatory risk profile.

of decreasing or offsetting 10 million tons of CO_2 by 2015 and will spend about $3 million annually to help fund projects to achieve that goal. However, absolute emissions will still increase. "This was an eye-opener for us, but it made us realize that mandatory, economy-wide action is necessary," says Duke's Leahy. "As a regulated utility, there is only so much we can do in a voluntary world, while balancing our mandate to provide least-cost electricity to our customers and a reasonable return for our shareholders."

Technological innovation is also needed. Duke continues to plan for its IGCC plant in Indiana and is proposing to build two 800 MW super-critical pulverized coal units in the Carolinas. Both projects were awarded federal tax credits as advanced clean coal technologies. Duke will include space for the inclusion of carbon capture technology at the sites, technology currently not commercially available. In addition, the geology in the Carolinas is not conducive for carbon sequestration—presenting another long-term technological challenge.

Like David Hone at Shell, Rogers worries how climate change could alter the fundamentals of his industry. "I worry that we are using 100 year-old technology. There will be a transformative technology. At what point will our generation and transmission lines become obsolete? There are a lot of things you might do, if you think there will be a new technology in 25 years. You need to hit your numbers with a short-term view, but you need to run your company with a long-term view." Having a seat at the policy table and influencing the final legislation will help ensure that it fits with Duke Energy's interests and future direction.

Staying One Step Ahead on Climate Change, Not Two

Swiss Re*

Where other companies in this book are motivated by the potential risk of future climate change regulation, Swiss Re stands out as being more at risk from the physical impacts of climate change itself. The insurance industry may experience dramatically increased costs due to a growth in climate-related effects; including growth in natural disasters, disease vectors and mortality rates over the next ten years.[112] But in keeping with the nature of reinsurance, the company has been working hard to integrate this risk into its business model. According to former Chief Executive Officer John Coomber (retired at the end of 2005), "While companies in most industries aim to avoid risks, reinsurers create value by analyzing risks and providing coverage for those they judge to be insurable."[113] "Climate change is a phenomenon that is starting to have a major impact on Swiss Re, its partners and clients. The question is no longer whether global warming is happening, but how it will affect our business, as well as our personal lives."[114]

Table 9

Swiss Re's Footprint (2005)

Headquarters:	Zurich, Switzerland
Premiums earned:	CHF 27.8 billion ($22.4 billion)
Employees:	8,882
Percentage of Emissions in Kyoto-Ratified Countries:	90 percent
Direct CO_2 Emissions:	6,829 MTons*
Indirect CO_2 Emissions**:	42,863 MTons
Aggregate CO_2 Emissions:	49,693 MTons
Target:	GHG Neutral by 2013
Year Target Set:	2003

* Metric tons.

** Measured as electricity use and business travel.

According to Chris Walker, Head of Sustainability Business Development, climate change is a central concern to the company because, "It could change the predictive model. If we don't have that model right, we could face problems in pricing some business going forward." In short, climate change undermines the fundamental model upon which reinsurance is based: that the earth's systems, though somewhat unpredictable in the short term, are stable in the long term. "What Swiss Re wants most is statistical regularity," says Brian Thomas, Manager of Content and in-house editor. But that statistical regularity is disappearing. According to Swiss Re, the insurance industry recorded around $40 billion weather-related natural catastrophe losses in 2004, the largest amount up until that date (see Figure 11). In 2005, the company estimates that total insured natural catastrophe property, and business interruption losses for the industry reached $78 billion. The total economic loss for windstorms in 2005 was estimated at approximately $209 billion.

Considering the company's substantial climate-related risk, Swiss Re has worked hard to promote understanding and action on climate change for more than a decade. Though the company received public notoriety for its early actions to address the issue, efforts at creating business opportunity through climate

* We would like to thank Nigel Baker, David Bresch, Pascal Dudle, Ivo Menzinger, Andreas Schlaepfer, Cosette Simon, Brian Thomas, Chris Walker and Mark Way for their contributions to this case study.

Figure 11

Rising **Natural Catastrophe Loss** Trends

Weather-related Natural Catastrophe insured losses 1970–2006 (property and business interruption)*

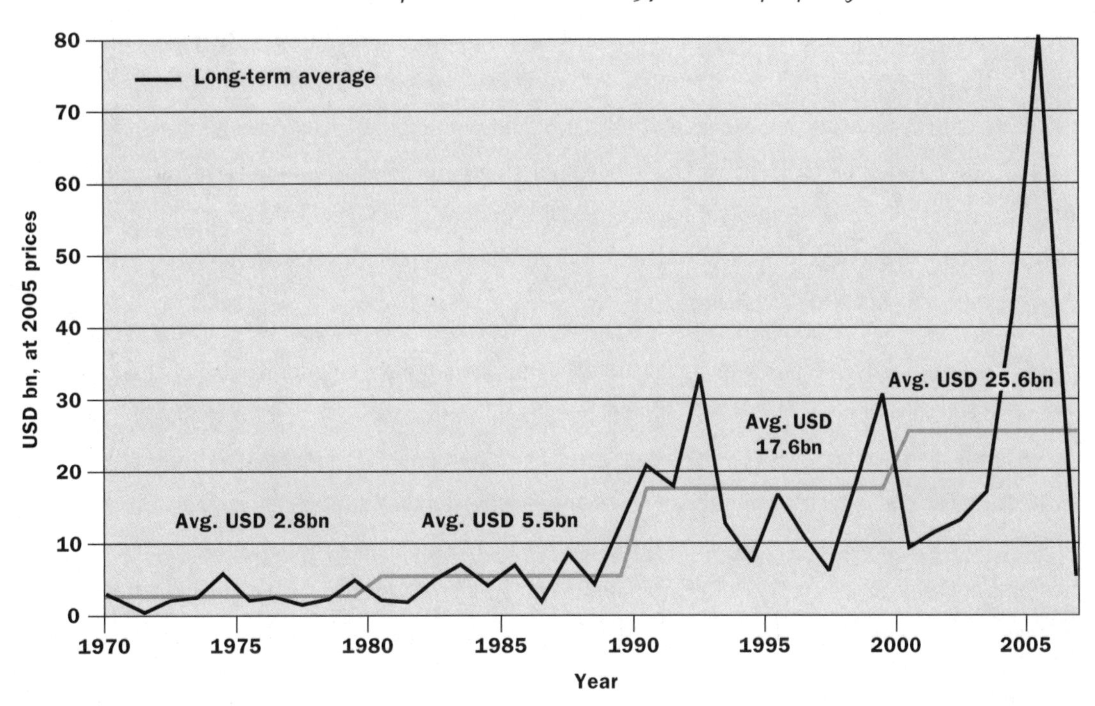

**Includes storm, flood, cold, drought, hail*

Source: Swiss Re sigma Catastrophe database

change mitigation-related products and services have fallen short of financial expectations. Walker feels that, "Considering how the political climate subsequently developed we were in the game too early; as a consequence, we lost momentum." Where Swiss Re was once the most visible financial services company on this issue,[115] that may no longer be the case. A 2005 *Business Week* article ranked the company eighth among its peers, behind climate newcomers such as HSBC and JP Morgan Chase.[116] Walker believes this is a result of a recent increase in attention to new entrants rather than a judgment of Swiss Re's 10 years of activity and commitment to the issue.

But the company is careful in its attempts to rectify this situation, wishing to be sure that there is no discrepancy between the company's external perception and internal reality. According to Mark Way, Head of Sustainability Issue Management and Reporting, "This is not about PR. We believe that the materiality of our commitment is comparable with the best of our peers. However, there is a danger to being perceived as a leader." David Bresch, Head of the Atmospheric Perils Group, agrees. "You should always remain one step ahead of the competition. But if you are two steps ahead, you lose the crowd. The ideal is for you to be the leader of the pack and everyone pulling in the same direction."

Company Overview

Headquartered in Zurich, Switzerland, Swiss Re is the world's largest reinsurer.[117] With operations in 70 offices spanning 30 countries, the company has three divisions: Products, Client Markets and Financial Services (which includes Asset Management). Forty-nine percent of Swiss Re's premiums come from North America, 38 percent from Europe and 13 percent from other parts of the world, mainly Asia. "We're a Swiss company with an American accent," quips Cosette Simon, Senior Vice President for Government Relations and Public Policy.

Swiss Re has historically operated as a quiet company in a low public profile industry. That said, there is a strong sense of pride within Swiss Re about its roles as a "knowledge company" and an "enabler" with a very long-term perspective. The company, for example, tries to avoid advocating a particular policy or regulation. Rather, it wants to be called upon for objective expertise on informing the development of that policy. "We want to be the first or second call someone makes if they want advice on the financial side of climate change," says Walker. As an "enabler," the company makes business deals and development projects possible by providing the necessary instruments to offset and diversify risk.

Because Swiss Re is a reinsurance company, it naturally tends toward more long-term and global perspectives when it comes to risk diversification. Climate change fits perfectly with that focus. According to Simon, "Climate change is a conservative issue with Swiss Re. It's about caretaking, stewardship, and a 50-year time horizon." As a result, Swiss Re has distinguished itself through a relatively long-standing strategy of external awareness building. In fact, more so than other companies in this book, awareness building, external outreach, and scientific research programs are as important to Swiss Re's climate-related strategy as innovative climate adaptation products such as the use of weather derivatives and catastrophe bonds, where the company is the market leader.

Climate Change Program Implementation

Swiss Re produced its first publication on climate change—*Global Warming, Elements of Risk*—in 1994. This book was ground breaking for the simple fact that it came from a financial services company and argued that the repercussions from climate change "could be enormous, with threats posed not only to citizens and enterprises, but also to whole cities and branches of the economy, even entire states and social systems."[118] With that as a starting point, the company has continued to establish its leadership position on this issue through efforts aimed at building awareness with clients and the broader public. Between 1995 and 1998, the company released four publications and conducted three client seminars on the topic.

Moving beyond education, Bruno Porro, former Chief Risk Officer (retired in 2004) tasked Walker in July 2000 to look at potential business opportunities related to climate change through a group-wide feasibility study. Walker identified nine areas of possible relevance to the company's business lines and, with the support of two executive board members on his advisory board, he identified nine champions within those areas who were willing to dedicate the extra time needed to explore them. In this process, Walker took special care to make sure that he brought the right people on board, remarking that he "only wanted intrinsically motivated managers—people who would read things at night."

Six months later, Walker presented his findings to the executive board. Not surprisingly, the central question following his presentation was, "Is this going to make money?" Though Walker admits that he did not have specific numbers to back up his rationale, he said that it would. The executive board supported the creation of Greenhouse Gas Risk Solutions (GHGRS) by approving a staff of four. "Before this," says Walker, "climate change was more a scientific concern. Now it was becoming more of a business development issue." The company narrowed the original list of nine areas to four business elements: investments, third party asset management, insurance/derivatives and emissions trading. Swiss Re's general approach to climate change is centered on the elements of research, products and services, management of its own emissions profile, and awareness building.

Research. Similar to DuPont, Swiss Re seeks to better understand its business exposure to climate change by developing internal expertise in climate science. According to Bresch, "The role of science at Swiss Re is to know what is possible because it helps the company identify, analyze, mitigate, and then, if possible, transfer our risk." Bresch also notes that while most reinsurers employ scientists, Swiss Re is unique in that it is one of the few that does all its modeling in-house. As a leading reinsurer, Swiss Re develops and maintains Natural Catastrophe (NatCat) state-of-the-art in-house models for all major perils worldwide, relying both on the knowledge and expertise of 30 NatCat experts in Zurich, Armonk (New York), Munich, and Hong Kong, as well as on active collaboration with leading scientific institutions worldwide.

The NatCat modeling did not always have such a prominent role in the company. As recently as 1980, Swiss Re employed only two full-time scientists within the NatCat unit. The staff grew over time to cope with the increasing complexity and the growing demand for detailed NatCat risk assessment and proper portfolio management. Starting from earthquake and windstorm models for key markets, the unit further developed tropical cyclone and flood models and now covers all relevant reinsurance markets worldwide.

While climate change has been monitored by climate specialists within the NatCat team since about 1990, quantitative analysis and integration in risk assessment and management processes started only when detailed impact studies became available. Where Swiss Re has been able to establish quantitative relationships, it has started to account for climate change risk in reinsurance pricing. This actually happened when Swiss Re decided to adjust its hurricane model in September 2005 to reflect the effects of natural climate variability, any superimposed human-induced trend, and increased modeling uncertainty. Swiss Re's NatCat experts follow and participate in actual research through collaboration with leading scientific institutions in order to identify climate effects at an early stage.

Products and Services. To be more proactive, Walker has adopted the mantra, "Distinguish ourselves relative to our peers." In that vein, he is searching for ways to improve underwriter's ability to bring climate change into policy decisions. "In Property and Loss (P&L), this is a stronger pitch," says Walker. "In Life and Health (L&H), it is harder."

One area where the company sees a possible link between its products and climate change is Directors and Officers coverage (D&O). According to Walker, "As soon as the obligation to reduce greenhouse gas (GHG) emissions becomes regulated, failure to comply or a mismanagement of carbon exposure could affect a company's performance and potentially create personal liabilities for directors and officers. Such regulations are already in force in some countries and are likely to become effective in the reasonably near future in the United States."

Signifying the risks of being a leader on climate change, Walker was once misquoted in the *Wall Street Journal Europe* as stating that Swiss Re will not provide coverage for climate change-related D&O risks. Instead, the company uses climate change as one measure among many to determine risk exposure. For example, with corporate clients, Swiss Re now looks to see if the applicant company has responded to inquiries from the Carbon Disclosure Project. If not, the company may add climate-related questions to its standard questionnaire to D&O insurance applicants.

The company is also testing the waters of integrating climate change-related risk factors into other traditional offerings. Business Interruption (BI) coverage is one promising area that Walker is evaluating. While BI insurance traditionally provides coverage for a plant that is forced to close temporarily, Walker is analyzing whether this coverage should include the value of tradable credits for ceased emissions during the shut down. In another area, the new Environment and Commodities unit has received a mandate to trade emissions. This unit is the combination of the old weather unit (insurance and derivatives) and emissions (SO_X/NO_X and GHG) and is presently in the process of staffing and has not yet started to trade.

Another important area of products and services is asset management, which had an investment portfolio of CHF 114.9 billion in company assets as of the end of 2005. Of these assets, 89 percent are invested in fixed-income, seven percent in equity, and the remaining four percent in alternative investments, including real estate. As early as 1996, Swiss Re asset management started to build up a dedicated Sustainability Portfolio comprised of investments which support sustainable development and efficient resource utilization. By 2004, approximately CHF 90 million had been invested in this area. In 2005, the company integrated the Sustainability Portfolio (including staff) into the alternative investments unit to benefit from a dedicated, institutionalized investment process.

Today, the company channels its sustainability investments into a number of sectors including alternative energy, water and waste management/recycling. More specifically, the company seeks opportunities representing medium to high risk-return profiles in: infrastructure investments such as wind-farm, biomass, and solar projects; investments in publicly quoted, small- to medium-capitalized growth companies; and cleantech venture capital investments, representing the highest risk-return profile. Lastly, the team seeks to invest in different geographical regions, with the target to reach a solid portfolio diversification in different markets. As tightening policy frameworks increase demand for such projects, the company's investment strategy is beginning to pay off. The portfolio's market value rose substantially in 2005 thanks to strong share performance as well as new investments.

Emissions Reduction. In October 2003, Swiss Re was the first company in the financial services industry to announce that it would eliminate or compensate for all of its GHG emissions, with a goal of becoming carbon neutral by 2013.[119] According to Walker, "We need to do this if we are going to be seen as credible."

At present, the company's GHG emission footprint is roughly 50 thousand metric tons, an amount management acknowledges is merely "a rounding error" of many of the companies in this survey. Direct emissions come from the combustion of office heating fuels (13 percent), and indirect sources include office electricity use (44 percent) and business travel (43 percent). Swiss Re plans to achieve a 15 percent reduction of these emissions through actual facility reductions and the remaining 85 percent through the World Bank Community Development Carbon Fund. The company is committed to increasing its purchase of renewable energy from 14 percent of the company's total worldwide energy consumption in 2005 to 37 percent in 2006 and 50 percent in 2007. Although the majority of this energy will come from wind, the particular source and quality in each location will depend upon regional availability.

Andreas Schlaepfer, Head of Internal Environmental Management, heads up the initiative and believes that for non-manufacturing companies like Swiss Re, substantial reductions in emissions resulting from energy conservation are quite easy; "If you've never focused on energy efficiency before, achieving 30 percent reduction is simple." However, achieving a 15 percent reduction will not be easy for Swiss Re since this target excludes savings made in previous projects. To achieve the company's goals, the program will focus on two primary areas: curbing emissions from Swiss Re's offices around the world and business travel.

Office emissions come from the nine buildings that the company owns and another 61 in which it rents space. While the nine owned buildings are responsible for 87 percent of the company's total energy consumption, the company includes rented office space in its carbon neutral initiative. Swiss Re employs a three-tiered approach to reduce its energy consumption. The first tier is zero-cost investments, such as turning down heating and cooling, and turning off lighting systems during non-working hours. The second tier focuses on small investments with paybacks of one year or less, such as motion sensors and compact fluorescent light-bulbs. The final tier includes refurbishments of property and buildings owned by Swiss Re, such as replacing cooling towers, generators, insulation or windows. The payback period for these investments can be as high as 10 years. Swiss Re has not established a formal budget to address these tiers, but will draw from the company's annual logistics budget.

To date, the company has conducted energy audits and provided recommendations for corrective measures in its three highest carbon-emitting offices. Based on the recommendations, local action plans have been drawn up for the next three years. Meanwhile, the company has learned some key insights into why some offices have more emissions than others. In some cases, it may just be age, location or that one building manager is more energy conscious than another. In other cases, operations that are split between two separate buildings with different property managers minimize the company's leverage with property management. Swiss Re is consolidating office space wherever possible and actively organizing tenant groups to create change within the management company.[120]

One prominent example of the company's efforts to become more energy efficient (and more visible) is its new office building at 30 St. Mary Axe in London. The building, known as the "Gerkin" after its unique shape, utilizes natural ventilation in addition to air conditioning. Due to this efficient design, it is expected that for much of the year the heating, ventilating and cooling (HVAC) systems can be switched off, thus reducing energy consumption and CO_2 emissions.

Emissions from business travel are the second, and more difficult, component of Swiss Re's carbon neutral initiative. Responsible for 43 percent of total company emissions, these emissions have been growing in both nominal and relative terms in recent years and are expected to overtake other emissions within the next two years. The reduction strategy is directed exclusively at reducing short-distance trips for internal meetings. According to Schlaepfer, it would be unrealistic and inconsistent with Swiss Re's business growth strategy to regulate business travel aimed at meeting current or potential clients, particularly in rapidly growing regions of the world such as Asia.

Although the company has not created any formal incentives encouraging employees to reduce internal travel, Schlaepfer has the support of top management for this initiative. In a CEO Newsletter in 2005, CEO Coomber touted the environmental and economic benefits from reducing business travel for internal meetings. Employees are required to secure the approval and signature of an immediate supervisor before taking such a trip. According to Schlaepfer, the most significant challenge is overcoming the human hurdle and unspoken professional incentive to network face-to-face with employees in other offices. To overcome this bias, Swiss Re provides employees with the latest telephone or video conferencing technology, and Schlaepfer arranges video conference training sessions to help mitigate any potential technological hurdle.

While the company plans to register its emissions reductions with the World Economic Forum's Greenhouse Gas registry, Schlaepfer states that Swiss Re will retire them rather than sell them. The company will also retire any renewable energy credits (RECS) that it purchases to meet a goal of 30 percent of its electricity purchases coming from green sources in the United States in 2006. The company may place carbon on the Chicago Climate Exchange (having joined in 2005) for trading purposes in the future, but initially the goal is to spur the market by demonstrating fungibility between markets by exchanging United States-based carbon credits with those from another jurisdiction. Explains Schlaepfer, "Our aim is to do something for the climate. This is a voluntary action, and it is not triggered by profit thinking."

Awareness Building and External Outreach

Of all the companies studied in this book, Swiss Re places the most emphasis on external awareness building within its climate-related strategy. The irony is that the company has historically sought to remain quiet and not draw attention to itself or its positions. Although the company had been in the United States for over 100 years, "no one knew who Swiss Re was," states Walker. "We were always a B2B[121] company." The company's approach to global warming ended this anonymity.

In July 2002, Swiss Re orchestrated a watershed event by sponsoring, along with AON, Duane Morris LLP and Natsource, a two-day conference at the New York Museum of Natural History called *Emissions Reductions: Main Street to Wall Street—The Climate in North America*. Bringing together more than 200 business, government and environmental leaders, this meeting was among the first instances in which Wall Street engaged on the climate change issue. More importantly for Swiss Re, it successfully garnered enormous attention from the press, public and financial community.

Building off this success, the company has continued to work to educate those within the financial industry. Using its in-house conference center (the Swiss Re Centre for Global Dialogue at Ruschlikon), the company creates forums to discuss "global risk issues and to facilitate new insight into future risk markets." The center has hosted three forums on climate change, including one in 2003 cosponsored by the International Emissions Trading Association (IETA) that focused on developing carbon markets.

In another ground breaking move, Swiss Re partnered with Stonehaven CCS Canada in 2003 to develop an educational video about climate change designed for the general public. *The Great Warming* is a three part (45 minute installments) television documentary highlighting the roots of climate change and its possible implications in the future. Narrated by singer/songwriter Alanis Morissette and actor Keanu Reeves, the internationally promoted series was filmed in eight countries on four continents and was endorsed by dozens of the world's leading scientists. First aired on Discovery Canada in 2004, it was subsequently broadcast in the fall of 2005 on the Public Broadcasting System (PBS) in the United States under the title *Global Warming: the Signs and the Science*. It has also been edited into a theater version which was shown in the spring and summer of 2006. In retrospect, *The Great Warming* is considered a huge success for the company. Not only did it distribute well to major television studios around the world, but Swiss Re has received only positive feedback on the final product.

From a more academic perspective, the company partnered with the United Nations Development Program (UNDP) and the Center for Human Heath and the Global Environment at the Harvard Medical School in 2004 to host a conference involving 250 scientific and business experts. They gathered to examine the physical and health risks of climate instability and formulate climate change scenarios and potential impacts on the environment, human health and the economy. Released in 2005, the final report—*Climate Change Futures: Health, Ecological and Economic Dimensions*—explains the links between climate change and human health.

Through such events and materials, the company has steadily transitioned from producing strictly client-centered publications to producing materials for a much broader public. It is clear that climate change has significantly altered the company's approach to external outreach. In Baker's estimation, "Climate change is one issue where we moved from internal to external dialogue." For example, according to Baker, "There was a lot of soul searching within the company on whether to get into television. Reinsurance is traditionally faceless." But he believes that *The Great Warming* is largely responsible for giving "a faceless Swiss company" some public recognition, particularly in the United States. "Although people may not know what we do, they know our name." Toward this end, CEO Coomber was instrumental in helping the company overcome internal concerns.

More importantly, the company's broader outreach approach is critical for creating awareness of climate change, and therefore its business interests. According to Baker, "Our clients' clients (such as you and me) are also a part of the problem. Reinsurers are at the back end of the game. Some insurers may be under less pressure to change the behavior of their clients because it doesn't hurt their pockets as much as ours. But we can't dictate. We must try to gradually build awareness." For example, *The Great Warming* was aimed largely at a North American audience, a demographic that the company believes is in critical need of climate change awareness building. As Gerry Lemcke, Deputy Manager of the Catastrophic Perils Unit, explains, "this TV series comes at the right time in the right country."[122]

Organizational Implementation

Within Swiss Re, some of the most significant future areas of group-wide strategic relevance are categorized as "top topics." Climate change has been a top topic since the program was developed in 2001. The selection process for top topics is currently run by the Issue Management Unit. Using sources such as SONAR (Systematic Observations of Notions Associated with Risk), this group of three employees conducts four or five meetings a year with staff to get input about relevant, emerging business issues. After a brief is drafted about each particular issue, it is submitted to the 15-member Issue Steering Committee, a diverse group of senior employees, up to executive board level, from areas such as property, casualty, and HR. It is the responsibility of this committee to determine whether a particular suggestion is categorized internally as an "issue," "topic," or "top topic," in increasing order of importance. Each categorization receives differing levels of attention and budgets. Beyond climate change, the other eight top topics, most of which come from the risk side of the business, are: natural catastrophes, water, insurance linked securities, liability regimes, mortality, nanotechnology, solvency and terrorism.

While top topics signify internal and external commitment to issues such as climate change, there are organizational challenges to generating internal consensus and support for the issue. Way argues that "internal awareness is built by making a clear link between climate change and our business bottom line." To increase awareness within the Asset Management Division, for example, the company has worked with Sustainability Asset Management (SAM) to bring in sustainability professionals to educate portfolio managers on how climate change, environmental, and social issues impact stock prices and the valuation process.

The company also works to educate employees on the changes they can make in their own lives to benefit the environment. And again, "You have to try and link the issue to employee's daily lives," says Schlaepfer. "Remind them to live up to Swiss Re standards (such as integrity) and take them home with you." Schlaepfer believes the company can do this because, in return, Swiss Re encourages employees to bring their diverse values and ways of thinking back to the company. But, says Schlaepfer, "the key is not telling employees what to do."

Educational efforts begin with new employees. In Zurich, a component of new employee orientation specifically focuses on climate change. In the last three years, the company has also held a series of marketing and educational initiatives during lunch hours to make the connection between climate change and employees lives more clear. The voluntary events focus on energy, business travel, commuting, video conferencing and other issues. In one event, the company arranged for 15 hybrid-electric vehicles to be brought to the Zurich office for all employees to test drive. The company has also arranged a series of "Lunch and Learn" sessions, during which internal and external speakers present climate change related research and information to all employees over lunch. The company recently organized an on-site climate change art exhibit that depicted glacial melting in various regions of the world by contrasting postcard images from the early 20th century with photo images from the early 21st century. Although all previous marketing efforts have been in the Zurich office, the company plans to repeat similar, tailored events in offices around the world. Most recently, the company instituted the "CO_{You2} reduce and gain" program in January 2007. This program supports employees' investments in personal measures that contribute to reducing greenhouse gas emissions, particularly in relation to mobility, heating, and electrical energy. Such measures, which vary according to regional circumstances and preferences, include low-emission hybrid cars, use of public transport, and the installation of solar panels or heat pumps. Until the end of 2011, Swiss Re will rebate each employee one half of the amount invested in these measures, up to a maximum per employee of CHF 5,000 or the equivalent in local currency.

Policy Perspectives

Swiss Re's foray into government relations is a relatively recent effort. In the United States, the company's operations are primarily regulated by individual states rather than the federal government. Cosette Simon, who joined the company in 2001, is the company's first government relations professional. (The company has recently hired a second person to work in this area.)

Although Simon admits that climate change issues account for less than five percent of her time, the company has leveraged its resources and has been vocal on Capitol Hill. Walker has traveled to Washington D.C. no less than 25 times and has testified before the Senate. In addition, Walker has testified before the New York state legislature's insurance committee, and the company has supported both California's Greenhouse Gas initiative by serving on the California Climate Change Advisory Board and the Regional Greenhouse Gas Initiative (RGGI), a cap-and-trade system covering seven states in the Northeast United States. These efforts are viewed by many within the company as an extension of the "knowledge company" mantra. As Simon explains, "We feel an obligation to share our expertise with policy makers. We search for the proper public policy, not just what is good for Swiss Re. We sometimes even engage in issues where we have expertise but we may not have a dog in the fight." She adds, this allows her to go to Congress "without an axe to grind."

When it comes to specific policy, the company is very open to suggestions. Given its vulnerability to the physical implications of climate change, what is most important to the company is progress of any kind. As Walker notes, Swiss Re has no vested interest in engaging conversations about "whether we need five percent or six percent reductions. We need 60 percent reductions to stabilize climate change. Period." According to Simon, "Swiss Re supported McCain-Lieberman because it is progress and has been the only game in town. We're just trying to get traction someplace. We may have supported something else if it had a chance of passage." In short, the company is willing to endorse policies in which the government is involved, compliance is mandatory, and, according to Simon, which include "market mechanisms that strike the right balance between environmental and societal policy objectives."

Swiss Re has also been involved in a number of global forums on climate change. For example, the company joined 23 multi-national companies in signing the declaration prepared by the G8 Climate Change Roundtable in 2005. The statement calls on the world's governments to create a long-term policy framework to allow for "clear, transparent, and consistent price signals" for carbon. In addition, the company has been participating for the last four years in discussions of the issue in the World Economic Forum in Davos, and has been closely involved in the climate change efforts of Prime Minister Tony Blair as part of the United Kingdom's leadership in the G8 process and elsewhere.

Moving Climate Change from the Periphery to the Core of the Organization

Like incubators at many companies, the dissolution of GHGRS was planned. When the group was formed, the intent was for it to serve as a center of competence on the emissions reduction issue and to look for business and investment opportunities for Swiss Re's existing areas of business. As such, it was to be a climate change knowledge facilitator for the company and was not intended to replicate existing business and investment functions. The concept was to develop lines of business in conformity with existing products and then pass them on to mainline offerings when the business was mature. With the emergence of the European ETS and Kyoto market mechanisms, the need for a separate unit was diminished as the Capital Markets and Advisory Units were convinced that the trading and derivative areas represented a complementary businesses opportunity to their existing weather business. As such, GHGRS, was successful in integrating the issue into various core businesses within the company, such as Capital Markets and Advisory (trading related products), risk awareness (D&O insurance) and Carbon/clean energy asset management (Conning).

Challenges Ahead

Given its early action on climate change, Swiss Re provides a wealth of lessons on how to act, as well as notable impediments that could be faced by a broad array of companies. Like other companies in this book, a key lesson from Swiss Re has been the importance of executive-level buy-in for the company's strategy. Walker says that whenever he approached senior management for support and resources for climate change activities, "I've always felt like I was knocking on open doors." According to Swiss Re's Menzinger, upper management's commitment to the issue is the most important factor ensuring that climate change remains a strategic area of focus. These comments were echoed by Baker who remarks that, in general, "Board level support silences internal opposition."

But recent events have altered the internal landscape at Swiss Re for moving forward. In the summer of 2005, Swiss Re was restructured just before a new CEO, Jacques Aigrain, was announced. Implications for on-going GHG initiatives include the creation of a centralized logistics department to oversee office space management and carbon neutrality. Also, the formal structure of GHGRS was dissolved. The group's mature offerings—including carbon trading and weather derivatives—were redistributed to mainline product groups. (See "Moving Climate Change from the Periphery to the Core of the Organization" on this page)

In addition to continuing his focus on D&O and BI insurance, Walker has been reassigned to act as a manager of Sustainability Business Development, which aims to bring other climate change and sustainability related products to market. Walker admits that "these efforts may not be huge potential revenue streams, but they will help to better manage risk both for clients and ourselves and integrate sustainability into the business and investment lines. This has benefits in technical knowledge and risk awareness, as well as leveraging the reputation of Swiss Re." More succinctly, Walker sees four areas of benefits for the company: leverage Swiss Re's knowledge of sustainability matters to generate high-quality investment opportunities and additional fee income; attract new clients to Conning Asset Management (Swiss Re's third party asset management company); provide superior risk-adjusted investment returns for its investors; and protect Swiss Re's brand by reinforcing its position as a leader in the corporate sustainability area, and avoiding association with "sustainability laggards."

Walker is also working to develop applications to assist companies to achieve carbon footprint neutrality. For example, the company tried partnering with the Commonwealth Bank of Australia to offer a GHG neutral initiative. This program offered companies three critical tools: a "platform for communicating climate change issues and a way to differentiate their products in a pre-regulated marketplace"[123]; a management system for calculating and obtaining the necessary offsets, as well as a system to manage potential compliance obligations; and a way to better enable the creation of carbon markets by allowing project developers the opportunity to monetize the environmental benefit that the project is providing. This program has evolved into the "Footprint Neutral" concept which the company is creating along with the UNDP to enable businesses, communities and consumers to voluntarily offset their footprints.

But, while initiatives like this hold promise for the company, limits in central coordination stand in the way. Although climate change is relevant for many departments within the company, "This is very much driven by individuals who have a commitment," says Menzinger. "The loss potential is enormous and our ability to diversify is limited. But there is also a huge opportunity in areas like weather derivatives. To get there, we need to make more internal consistency and coordinate our efforts on climate change better across the business as a whole." Like other companies, Swiss Re is challenged by the task of directly linking climate change to the balance sheet. Says Baker, "A lot of research is lacking on modeling to connect the science and economics." Dudle wants to see more work in "getting the numbers to make the business argument for investment decisions."

But in the end, Swiss Re remains persistent. Due to the nature of climate change and the impact it can have on the company's business model, it is imperative that Swiss Re continue on the path it began in the early 1990's to build more robust bridges between science, human health, environment, and economics on the climate change issue. At times, the company has gotten too far ahead on the issue. Walker, for example, expresses disappointment at the lost opportunities when the market for carbon collapsed with the Bush administration's refusal to ratify Kyoto. "We got in too early and lost some momentum." But, he adds, "As a change agent, you have to be willing to take your lumps. Luckily, as a reinsurer, we're patient. Now it is easier to make the business case." Recent policy developments on climate change around the world are leading to greater opportunities for the company's efforts. According to Simon, "I've seen a real change in the last 12 months. I'm sensing a real shift." So, while the company may have gotten too far ahead at times, Menzinger believes that its early approach paid off because "it moved the market and raised the company's profile."

Shifting from Risk Management to Business Opportunity

DuPont*

Once again, DuPont is transforming itself. One of the oldest companies in the United States, DuPont began as a black powder[124] company in 1802; transformed into an explosives manufacturer in 1880; turned to polymers, paint, plastics and dyes in the early 1900's; added energy to its portfolio in 1981; and now, as it enters its third century, is pursuing new business lines of agriculture, nutrition and bio-based materials.[125] To make this latest transition, the company has been shifting away from lower-growth businesses that are heavily reliant on fossil fuels—evidenced by the sale of the Dacron®, Lycra® and Nylon® divisions in the early 2000's—and expanding into high-growth businesses such as bio-based materials—evidenced by the acquisition of Solae[126] and Pioneer Hi-bred International[127] in 1999.

Table 10

DuPont's Footprint (2005)

Headquarters:	Wilmington, DE
Revenues:	$26.6 billion
Employees:	60,000
Percentage of Emissions In Kyoto Ratified Countries:	8 percent
Direct CO_2e Emissions:	9.64 MMtons*
Indirect CO_2e Emissions**:	4.02 MMtons
Aggregate CO_2e Emissions:	13.66 MMtons
Target:	65 percent reduction in GHG below 1990 levels by 2010
Year Target Set:	1994 Recast in 1999

* Million metric tons.

** Measured as purchased electricity & steam.

But at present, DuPont is still the 2nd largest chemical manufacturer in the United States, and remains heavily dependent on fossil fuels for energy and feedstock in its industrial chemicals, polymers, and high-performance materials businesses. As such, climate change is an issue that the company cannot, and does not, ignore. In 2005, DuPont was listed as the "top company of the decade" (1995-2005) by *Business Week* magazine[128] and Ceres picked the company as the leader in its industry;[129] both awards are based on accomplishments in greenhouse gas (GHG) reductions. But "DuPonters" (a name that employees use in reference to themselves) still see a pressing need to do more. In fact, the challenge they now face is the most important—transitioning their company's treatment of climate change from one of risk management to one of business opportunity. Don Johnson, Group Vice President (VP) for Operations and Engineering, says, "We have to begin to think of energy as a value and not as a cost." James Porter, VP of Safety, Health, and Environment and Engineering, adds that, "to shift from risk management to business opportunity you need to understand the value chain. You've got to discover new ways to use what you've got, while also developing new materials to serve new needs and concerns."

Company Profile

Based in Wilmington Delaware, DuPont has operations in more than 70 countries, 60,000 employees worldwide and 2005 revenues of $26.6 billion. The company's products and services span agriculture, nutrition, electronics, communications, safety and protection, home and construction, transportation and apparel.

* We would like to thank John Carberry, Uma Chowdhry, John DeRuyter, Linda Fisher, Craig Heinrich, Don Johnson, Mack McFarland, Ed Mongan, Michael Parr, James Porter and Dawn Rittenhouse for their contributions to this case study.

DuPont's corporate vision is "to be the world's most dynamic science company, creating sustainable solutions essential to a better, safer and healthier life for people everywhere."[130]

In fact, safety has always been a key component of DuPont's culture, stemming from the dangerous nature of the company's first product, black powder. Porter states that with respect to safety, health and environment, there is a "cultural bias to do the right thing." But it is DuPont's long history of scientific innovation that is at the center of the organization. With more than 75 research and development (R&D) and customer service labs,[131] the company uses integrated science to develop new products and vigorously pursue what it terms "knowledge intensity"—getting paid for what the company knows rather than simply for what it makes.[132]

DuPont prides itself on being at the forefront of the environmental sustainability movement, a leader in ozone layer protection (DuPont was awarded the 2002 National Medal of Technology for "CFC Policy and Technology Leadership"), and an early actor on climate change. DuPont's sustainable growth initiative is the latest evolution of strong CEO leadership on environmental issues. Former CEO Dick Heckert (1986-1989) led the decision to phase out fully halogenated chlorofluorocarbons (CFCs) in the late 1980s. Former CEO Ed Woolard (1989 -1995) referred to himself as the "Chief Environmental Officer" and set the company on a "goal of zero"—zero injuries, illnesses, incidents, wastes and emissions. And present CEO Chad Holliday, former chairman of the World Business Council for Sustainable Development and co-author of the sustainability book *Walking the Talk*, set sustainable growth goals for DuPont which require an integration of economic, social and environmental performance.

But there is more to these environmental efforts than just top-down leadership. A reinforcing loop is at work—strong leadership is born out of the committed culture, and in turn relies on the culture to set and achieve aggressive goals for initiatives. The company's strong, goal-oriented culture "drives everything," according to Ed Mongan, Global Manager for Energy and Environment. "We set goals and everyone feels challenged to do their part. We openly track progress by individual sites and business units to meet those goals so no one can hide." The key to setting goals on environmental issues is strong and forward-looking leadership; the key to achieving the goals is the corporate culture.

Climate Change Program Implementation

DuPont's actions related to climate change were foreshadowed by its experience with ozone depletion in the 1970s and 1980s. Relying on its strong scientific expertise, the company reacted to the ozone issue when it first emerged in the scientific journals. According to atmospheric scientist and DuPont Environmental Fellow Mack McFarland, Molina and Roland's 1974 *Nature* article linking CFCs with ozone depletion "got the ball rolling." As the largest manufacturer of CFCs at the time, DuPont initiated an internal task force to address the issue and senior management was briefed. Realizing that regulation was imminent, DuPont began exploring alternatives. In March 1988, after the signing of the Montreal Protocol, DuPont announced a voluntary and unilateral phase-out of CFCs through an orderly transition to alternatives. In 1991, the company began operation of the world's first manufacturing facility for the hydrochlorofluorocarbon HFC-134a, an alternative to CFCs. Today, CFC alternatives comprise two to three percent of DuPont's portfolio.

This experience taught DuPont that understanding atmospheric science, engaging in the policy arena, and realizing the market impact of future regulation was critical for its future growth. As *Business Week* describes it, DuPont is "an experienced hand at making the most out of changing regulations."[133] When the Intergovernmental Panel on Climate Change (IPCC) issued its first assessment report in 1990, DuPont saw a familiar scenario playing out and, given its experience with CFCs, then-CEO Woolard directed that DuPont become an early adopter of a GHG reduction strategy.

The company began measuring and tracking their largest GHG emissions—CO_2, nitrous oxide (N_2O) and HFC-23—in 1991 and also made an internal commitment to reduce net emissions. This action coincided with a larger expansion of environmental efforts at DuPont. In 1992, the company published its first external environmental report and an Environmental Policy Committee was created on the Board of Directors.

DuPont made its internal commitments to reduce GHGs and energy use (per pound of product) public in 1994 by becoming the first company to join the Environmental Protection Agency (EPA)/ Department of Energy (DOE) Climate Wise program. The initial goal was to reduce GHG emissions 40 percent below 1990 levels by the year 2000. Establishing the goals was a two step process. First, each business unit identified possible reductions. Then, the Safety, Health and Environment Excellence Center (a Corporate function comprised of policy and technical experts under the VP for Safety, Health and Environment, the role of which is to develop and facilitate implementation of corporate environmental policy) pushed those reductions further, creating a stretch goal.

The first actions taken toward achieving the GHG reduction goals were aimed at the "low hanging fruit" in the company's operations. At the time, there was little sense of opportunity for competitive advantage other than getting ahead of the curve on regulation. DuPont's "low hanging fruit" consisted of reducing emissions of two potent GHGs: N_2O, with a Global Warming Potential (GWP) of 310 times that of CO_2, and HFC-23, with a GWP value of 11,700. In fact, given these high GWPs, CO_2 emissions were not a major issue for the company when GHG reduction goals were first initiated.

In 1991, a scientific paper[134] implicated Nylon production as a source of atmospheric N_2O, a GHG regulated under the Kyoto Protocol. In response, N_2O producers reached an industry-wide agreement in 1993 to reduce emissions by 1999.[135] To reach this goal, DuPont developed an end-of-pipe capture and destroy technique which eliminated 90 percent of emissions at a cost of $50 million with no payback to the business unit's profit and loss (P&L) statement. This additional burden was acknowledged by headquarters and earnings expectations for the unit were adjusted accordingly. For DuPont, accepting the $50 million hit was not only an issue of avoiding government regulation, but also of sticking to the company's principles by "doing the right thing." DuPont shared the technology with the other N_2O producers in the agreement as it was an end-of-the-pipe addition, separate from the core process, and substantial benefits required adoption by the entire industry.

The second target GHG, HFC-23, is an unintended byproduct from the production of HCFC-22, a common refrigerant, and part of DuPont's product line.[136] Reductions of HFC-23 were primarily achieved through a process improvement, resulting in greater yield of HCFC-22 and therefore reduced HFC-23 byproduct. Additional reductions were accomplished through thermal destruction of all or a portion of the remaining HFC-23. Unlike the N_2O reduction technology, the HFC-23 reduction was not driven by an industry-wide agreement and involved an alteration in the core process that resulted in competitive cost savings. Therefore, the technology remained proprietary.

When it was realized that the initial GHG reduction goals would be readily achieved through these two initiatives, DuPont management moved swiftly to establish new goals. The new targets, set in 1999, were expanded to incorporate energy efficiency goals and to fit with DuPont's sustainable growth initiative. They consist of three elements: hold energy flat at the 1990 baseline; source 10 percent of energy from renewable sources at cost competitive rates; and reduce net GHG emissions to 65 percent below 1990 levels, all by the year 2010. An additional, more far-reaching and market-facing goal is to create at least $2 billion of revenues from products that create energy efficiency and/or significant greenhouse gas reductions from customers by 2015. Maintaining the 1990 baseline for the GHG reduction goal was a deliberate move, consistent with the baseline for countries under the United Nations Framework Convention on Climate Change and also reflective of the company's desired baseline for early action credits.

To achieve these new goals, "We have to attack energy," says Linda Fisher, VP and Chief Sustainability Officer. "We have a heavy dependence on fuel, and so rising energy prices are a major concern." DuPont is vulnerable to energy prices on two fronts because much of the feedstock it uses is derived from hydrocarbons, especially natural gas. This vulnerability was reflected in DuPont's fourth quarter 2005 earnings, which were half the amount predicted due to higher energy and ingredient costs, as well as hurricane disruptions, plant outages and lower sales in some segments.[137] Uma Chowdhry, VP of Central Research and Development, states it simply: "What energy prices have done to us focuses the mind very quickly."

DuPont's attention to energy efficiency is currently at a point of transition. According to John Carberry, Director of Environmental Technology, energy efficiency efforts between 1990 and 2000 were dominated by yield, capacity and utilization gains; cogeneration and power partnering; and replacing low value/high energy products with those that are high value/low energy. For example, coatings for the auto industry are being replaced with very low Volatile Organic Compound (VOC) coatings, and commodity fibers are being replaced by Pioneer HiBred's corn and soy seeds. Since 2000, he says the focus has been more fine tuned and aimed at instrumentation changes to affect yield, capacity and utilization; process changes; continuing use of combined heat and power; and modern heat management including insulation, steam traps, waste heat recovery and modern motors. The difference between the past and the future is that the latter is highly investment intensive.

Through the company's efforts, energy use has decreased six percent compared to 1990 levels, despite a 40 percent production increase, saving the company over $3 billion since 1990 and yielding a decrease in GHG emissions of 420 million metric tons. This financial savings figure is calculated as the costs avoided through energy reductions achieved by improving yields and creating less energy-intensive product portfolios versus the business as usual scenario.

Sourcing renewable energy, the second energy goal, has the potential to reduce upstream emissions, fuel costs and exposure to volatile price fluctuations. While progress in this area has led to an annual cost savings of approximately $8 million, meeting the goal of 10 percent has proven challenging. According to Porter, this will be the "toughest goal, yet if we didn't set a goal, we wouldn't have done anything." Cost-competitive projects are relatively scarce and difficult to identify. The company has only been able to source about five percent of its energy from renewable sources, with most efforts coming from the use of landfill gas. In one example, the company partnered with a municipal landfill near its De Lisle, Mississippi plant. A third party laid seven miles of pipeline and installed compression equipment to bring low-cost gas for the plant's boilers. Although it is a less reliable source than the local gas provider, the effort has displaced 30 to 50 percent of natural gas used to run the boilers.

With regard to the third goal of GHG emission reductions, DuPont has been quite successful. As of 2003, DuPont achieved a 72 percent reduction from 1990 GHG emissions. After the 2004 divestment of the nylon business, Invista®[138] related GHG emissions were removed both from the baseline and the realized reductions and overall reductions were recalculated as 60 percent. (This practice of recalculating emissions follows the WRI/WBCSD GHG protocol as well as that of the Chicago Climate Exchange).

As the company's programs have developed, its strategies have become more sophisticated. Going forward, the challenge for DuPont is to treat climate change and energy efficiency as business opportunities by connecting them to the overall objectives of the firm. Company leadership believes that the right product mix will offer an advantage in a carbon-constrained world. Fisher, who is tasked with embedding sustainable growth into strategic planning, gives her view on what climate change means at DuPont, "It's more than just science. It is also a matter of understanding our role in both the problem itself and our opportunities to address it; and to get internal agreement on that."

For DuPont, the business aspect of the issue has two components: risk management—will DuPont be put at a competitive disadvantage from carbon constraints?—and business opportunity—can DuPont capitalize on carbon constraints to expose new market opportunities? According to Fisher, "In developing future business plans and strategies, we need to understand the implications of GHG restraints and whether they pose a risk or opportunity for our family of products." As regulation becomes more likely, such analyses will be further developed.

John Ranieri, VP and General Manager of the Bio-Based Materials division, sees a number of areas in which DuPont has developed "sustainable innovations" that have already shown great promise.[139] "The real challenge is beyond our own footprint, it is in the market opportunities," says Fisher. "Can we measure the benefits to the customers? Are there growth opportunities? Some businesses are doing it. We need to work closely with customers to identify their needs and work to find a solution for them," either from new uses of old material or from developing new solutions to customer problems. Since 2000, DuPont has steadily increased its revenue from new products, growing from 20 percent of revenue from products introduced over the previous five years in the early 1990's to 34 percent in 2006.

For example, customers in the auto industry required coatings with much lower VOC than previously available, which, once developed, required much less organic solvents from the company's suppliers. Also, DuPont developed a special grade of Tyvek® house wrap[140] in response to European customers (where residential reductions are part of the national climate strategy) for a product that would lower CO_2 emissions and heating bills. In some cases, DuPont engineers work with customers to help them reduce their own energy use, delivering higher value to customers and ultimately enhancing business through closer customer relationships and a stronger understanding of customer needs. Such efforts have been rewarded by larger or longer-term contracts.

Looking forward, DuPont has identified the most promising growth markets in the use of biomass feedstocks that, through metabolic engineering, can be used to create new materials such as polymers, fuels and chemicals, new applied BioSurfaces in the personal care, coatings and colors areas and new Biomedical materials for use in the cardiovascular and dental fields. The company has set a goal to nearly double its revenues from such non-depletable resources to at least $8 billion by 2015.

One promising development is Sorona® polymer. In a joint venture between DuPont and Tate & Lyle PLC went on-line in the fourth quarter of 2006, with the first shipments of 1,3-propanediol beginning in January 2007. Bio-PDO is the key building block for the new polymer using a proprietary fermentation and purification process based on corn sugar. This bio-based method uses less energy, reduces emissions and employs renewable resources instead of traditional petrochemical processes.

Another promising development is the 2006 creation of a partnership with BP to develop, produce and market a next generation of biofuels. The two companies have been working together since 2003 to develop materials that will overcome the limitations of existing biofuels. The first product to market will be biobutanol, which is targeted for introduction in 2007 in the U.K. as a gasoline bio-component. This biofuel offers better fuel economy than gasoline-ethanol blends and has a higher tolerance to water contamination than ethanol.[141]

Both of these developments represent the new direction in which the company is headed—one that significantly reduces the company's environmental footprint. According to Chowdhry, this is not a subtle shift, but rather a significant change in product lines and research focus for DuPont. She is hoping that DuPont will soon be known for leading the industrial biotechnology revolution and predicts that over 60 percent of DuPont's business will stem from the use of biology to reduce fossil fuels in the next few decades.

Organizational Integration

To integrate climate-related strategies into the business, DuPont employs a vast network of teams and committees. Overseeing and driving this complex structure is strong leadership from the top. CEOs Holliday and Woolard are (and were) both visionary spokesmen for the company's goals on environmental issues and personally involved in pushing the company to achieve that vision. Mongan describes one pivotal moment, "We almost missed our 2000 goal. One business said it was too expensive. The CEO and Paul Tebo (former VP of Safety

Health and Environment from 1993 to 2004) sat down with the business manager and firmly stated, 'we will not miss this goal!'" That kind of personal attention to the issue leads Mongan (and many others) to list the most important ingredient in initial successes on climate change as the "CEO staying the course."

Beyond strong leadership, achievement of the goals is encouraged and diffused in several ways. First, goal setting involves a broad spectrum of representatives throughout the company. This is an effective way to create buy-in for the climate-related strategies.

Second, while attaining individual goals is left largely up to the business units, their progress is tracked through the Corporate Environmental Plan (CEP), a database that captures environmental performance (such as waste, emissions, GHGs, and energy) annually from global facilities and tracks future reductions or increases in alignment with business plans. It is maintained and managed by the corporate Environment and Sustainable Growth Center (a Corporate function comprised of policy and technical experts under the VP and Chief Sustainability Officer whose role is to lead the development and facilitate implementation of corporate sustainable growth programs and policies.)

Third, Sustainable Growth Reviews, performed by the Environment and Sustainable Growth Center, provide an opportunity to discuss challenges and opportunities within specific business units. In these reviews, experts from the sustainable growth team meet with business leaders annually to review key performance indicators for safety, health, environment and sustainability in relation to business and corporate commitments and goals. The discussion focuses on how these goals and indicators are integrated into their business plans and strategies, especially with regard to future growth plans and opportunities.

Fourth, DuPont ensures organizational buy-in and action on its climate-related strategies by linking compensation and bonuses for key employees, such as business leaders and energy experts, to program results. This provides an incentive, but remains a small portion of overall compensation for these individuals.

Finally, local champions are a critical factor for both programmatic and cultural reasons in an organization with decentralized businesses such as DuPont. That is why DuPont created Competence Centers to operationalize its goals. For example, energy experts within each business unit combine to create the Energy Competence Center, a formal network of energy professionals. Their job is to incorporate the ideals of energy efficiency into the operations of DuPont by embedding climate issues into decision making, examining the entire value chain, and involving individuals wherever possible (see "An Energy Efficiency Champion on the Ground" on page 97).

In an organization that depends upon a common culture to achieve buy-in for new initiatives, communicating the importance of climate change is vital. One way in which DuPont achieves recognition is through the Sustainable Growth Excellence Awards, where environmental projects in business units are submitted for corporate review. Of the 400 or so projects submitted every year, 12 finalists are chosen, rewarded with a dinner with the CEO, recognition throughout the company, and $5000 to donate to the charity of their choice. Many of the project examples mentioned in this case study were previous award winners.

An Energy Efficiency Champion on the Ground

One energy champion in DuPont is Craig Heinrich, leader of the global energy team for Titanium Technologies; a fast-growing division with plans to double production by 2010 from 1990 levels while increasing energy use by only 40 percent. This is no small task given that energy comprises a significant percentage of the selling price of titanium dioxide (TiO_2). Heinrich must be a vigilant internal salesman, aware of everything going on in his department. In describing his job, he states, "You need to communicate, you need to network.... The business case for energy efficiency has grown increasingly strong as energy prices have escalated," says Heinrich. "Even so, we have discovered the value of having an advocate for continued emphasis on improved energy efficiency. That is the role I play. It is necessary to repeatedly communicate the value so projects receive the appropriate priority."

One method he employs to stay ahead of new projects is Sustainability Screening, a process which evaluates a program's energy consumption and GHG emissions as part of the capital authorization process. The screening is performed early in the process, prior to other review steps, and involves both business unit and corporate level personnel. His Energy Competency Center's efforts have led to approximately 10 percent of the business unit's capital budget being invested in programs that improve energy efficiency, bringing year-over-year savings of $3 to $5 million. According to Heinrich, some projects may have a return of 300 to 400 percent. "For example, an air cooled condenser was used to supply desuperheating water[142] to one of our plants. We are switching to a third-party supplied reverse osmosis system, improving energy efficiency and reducing water costs." By outsourcing the project, DuPont avoided the capital costs.

Heinrich's goal is to incorporate energy efficiency in every project possible. As he describes his projects, very few of them are exclusively energy, often having an aspect of quality, volume or other emissions. But because large capital investments are being made to facilitate business growth, Heinrich has the opportunity to add energy and environmental improvements up front, before investment occurs. "Energy efficiency needs to be integral to the process. It cannot be an add-on," he states. But in units where energy projects are set to compete for limited resources against more mainstream investment proposals, the challenge is greater.

Reducing energy consumption in capital investments can often be met with resistance, particularly if the pool of resources is dwindling. John Carberry points out that the certainty of returns in energy efficiency projects can actually become a liability. The company has ruled out such instruments as lowered hurdle rates, internal carbon shadow pricing or a set budget for energy efficiency projects. "Energy efficiency must meet the same hurdle rate as other projects. The problem is that when we pitch 20 percent return with 99 percent certainty on energy, we lose to a marketing group pitch of 40 percent return with 60 percent certainty."

But, while energy efficiency projects are required to be cost competitive, and compete with all other capital projects for funding, many environmental projects, including those within the sustainable growth and climate change initiatives, are done with no capital return on the justification of either avoiding potential regulatory or legal liability, or avoiding reputational damage. The distinction is between projects that deal with risk management and projects that present a business opportunity.

As for the aggressive growth and efficiency goals set for Heinrich's unit, he prefers it that way. "You need the tension of a very challenging goal. Inspirational goals call an organization to act beyond conventional boundaries. These goals are built on the premise that real potential is beyond our ability to envision. An easy goal fails to challenge the creative potential of the organization." His advice for any company undertaking a climate change program is to "get passionate people engaged and challenge them to do something really extraordinary. They need a vision beyond what they can perceive and they need leadership to get them excited about what they can achieve."

External Outreach

As with other companies in this book, DuPont engages a number of stakeholders, including civil society, customers, trade associations and government. Managing these relationships helps DuPont build knowledge, convey actions and concerns, understand trends and engage more effectively in the political arena. Maintaining open communication channels enhances the company's business in the long run.

DuPont currently engages with non-governmental organizations (NGOs) in a multitude of ways. Often, a partnership is formed to meet specific project goals, with the primary driver being the expertise and different points of view brought to the table by NGOs. According to Fisher, "You can learn a lot from NGOs. They can open your eyes to market opportunities. Also, they add legitimacy to our environmental commitments. A big branded corporation stating its efforts sounds like public relations, but an NGO recognizing them carries a lot of weight, both internally for employees who are passionate on the subject and externally." Examples include partnering with the World Resources Institute and its Green Power Market Development Group to assist in meeting the 10 percent renewable energy goal, and joining with the Pew Center as a member of its Business Environment Leadership Council.

Unlike some other companies in this book, one venue for DuPont's external outreach on the climate change issue has been through the sales and marketing departments. As publicity surrounding DuPont's leadership in climate change initiatives increases, and general awareness of these issues grows, customers have been calling upon DuPont to deliver new, better performing products that are relevant within a carbon-constrained world.

In an example of collaborative partnerships, DuPont is leading a four-year, $38 million consortium (with NREL, Diversa Corporation, Michigan State University, and Deere & Co.) to develop an "Integrated Corn Bio Refinery." With $19 million in matching funds from the U.S. DOE, the consortium will design and demonstrate the feasibility of the world's first fully integrated bio-refinery, which will be capable of producing a range of products from a variety of plant-material feedstock; for example, converting corn into bio-derived chemicals, like Bio-PDO™, and bio-fuels, like ethanol. It "will create a new business model for sustainable production of chemicals, fuels and energy," says CEO Holliday.[143] "The technology will lower reliance on petroleum, reduce greenhouse gases, and create a global and sustainable bio-based economy."

DuPont is also a member of numerous trade associations, including the American Chemistry Council (ACC), the International Climate Change Partnership (ICCP), and the Council of Industrial Boiler Owners (CIBO). DuPont's involvement in these organizations represents the full gamut of industry issues, and the company works within these organizations to further climate change issues. Its efforts are more or less aggressive depending on the particular organization. According to John DeRuyter, Principal Consultant, Energy Engineering, "You should not become overly aggressive if you cannot get agreement. And with the ACC it can be very hard to get agreement with companies on either end of the spectrum." Recognizing that the diverse set of companies within associations do not always share their views, DuPont takes a cooperative approach, focusing its climate change efforts within organizations that are actively engaging the climate issue, like the Pew Center, the ICCP and the Business Roundtable.

Frustration with the Clean Development Mechanism

John Carberry calls the Clean Development Mechanism (CDM) "brutally political and complex" and like others in this book, feels that it is not living up to its potential. Mack McFarland believes that the principles of CDM are correct but the implementation rules need to be fine tuned. For example, he explains, under the present rules, "HFC-23 destruction [a waste byproduct] can be worth more than HCFC-22 production [a commercial product]!"

He explains, "A make-rate of 4 percent (the percentage of byproduct HFC produced) is the default value in the IPCC inventory guidelines for countries to use in plants where HFC-23 byproduct is not measured. When measured and managed, the lowest make-rate is normally just over 2 percent." Using proprietary technology in DuPont's Louisville plant, a make-rate of 1.37 percent was achieved, resulting in more of the desired product and less waste byproduct. The process was not expensive, but has effectively reduced the amount of HFC-23 produced. Yet three (non-DuPont) facilities with approved CDM projects are producing HFC-23 at nearly three percent.

Given that Certified Emissions Reductions (CERs) are selling at a price of about $10/ton of CO_2e, one could make more money from selling the CERs resulting from the destruction of the HFC-23 than they could selling the intended product (approximately $3.50 for destroying the HFC-23 associated with production of one kilogram of HCFC-22 that was selling in some regions for around $1.80). The originally approved methodology has since been modified and would allow credit for destruction of HFC-23 up to 4 percent, providing revenue of $4.70 for the destruction of HFC-23 associated with production of one kilogram of HCFC-22. This, in effect, rewards operations for being less efficient.

DuPont supports inclusion of HFC-23 projects under CDM but believes that CDM should not provide incentives that discourage use of Best Available Technology. The financial incentive described above would have encouraged new plants to make as much HFC-23 as possible up to 4 percent rather than optimizing the process to make as little HFC-23 as possible. The subject of the methodology for HFC-23 CDM projects for new plants is currently under discussion by the Parties to the Kyoto Protocol and UNFCCC.

McFarland concludes, "DuPont submitted comments under the CDM process on this issue. But right now CDM discourages the use of the Best Available Technology for reducing HFC-23 production in the manufacture of HCFC-22." Justifying DuPont's actions despite the CDM problems, he concludes, "We would not have looked for such a solution to reduce the amount of HFC-23 produced if not for the internal commitment to climate change and the need to meet that commitment on the most cost-effective basis."

In spite of all these initiatives, DuPont has minimal engagement with its shareholders and the broader investment community on climate change. Instead, the company's efforts on climate change are helping it avoid shareholder action. "We have not had to respond to proxy resolutions because of our proactive actions on the issue," reports Mongan. According to Fisher, "Mainstream institutional investors are not as focused on this issue in the United States as they might be. That could all change if legislation is enacted."

Policy Perspectives

As with trade associations, DuPont has taken a cooperative approach to engaging government on the climate change issue. In the 1990's, DuPont consulted with the Clinton administration and Capitol Hill representatives regularly. The company was quite active in the development process for the Kyoto Protocol, advocating market-based systems that shift capital to the most cost effective solutions; such as the Clean Development Mechanism

(CDM), a program that has frustrated the company thus far (see "Frustration with the Clean Development Mechanism" on page 99). DuPont has played an active role in advising and commenting on the development of the E.U. ETS. DuPont was also very active in the development of the U.K. ETS and participated in registration and trading of U.K. allowances through its Invista® subsidiary (now divested). Because climate change is a global problem, a global solution that includes all industrialized countries is critical.

Fisher believes that participation on the part of DuPont and other companies in domestic policy development is vital. "It is important for industry to help government find cost-effective solutions to the climate issue," she explains. "Government can't do it alone. They don't have the capacity to understand all the implications of the different policy options. The public comment period provides the government with critically valuable information." More recently, Fisher describes DuPont as "somewhat engaged, but not high profile" on government lobbying related to climate. "It takes resources to lobby and, as Congressional action on this issue gets more intense, we will put more time and energy into it."

Lobbying efforts dropped off when it became clear that the United States would not take action on climate change. Time and resources were spent on more critical issues, such as natural gas prices and availability. But renewed interest from policy-makers has DuPont stepping up its activity. Today, the company is struggling with the balance between the desire to see movement toward a federal standard with credit for early action, and the concern that comes from not wanting to alienate or adversely affect its customers by advocating aggressive actions.

Looking toward future regulation, DuPont sees an opportunity for longer-term views that encompass a global system with developing countries, including China and India. "This is an ideal time for renewed U.S. leadership on the issue," says Michael Parr, Senior Manager of Government Affairs. "We won't see China and India on board while the U.S. is on the sidelines."

Towards that end, in January 2007, DuPont joined with other leading U.S.-based corporations and environmental leaders to call upon the U.S. federal government to act on climate change legislation as a member of the United States Climate Action Partnership (USCAP). In February 2007, DuPont also endorsed the Global Roundtable on Climate Change, stating that they strongly believe in the principles outlined by the Global Roundtable on Climate Change and encourage other organizations to support this effort to advance action to address this critical issue.

One of the most important, if not *the* most important, aspects of policy for DuPont is recognition of early voluntary action. Whether these early actions are an asset or a liability depends on the baseline set in the final form of regulation. The other major critical issue is the effect of legislation on natural gas prices, an important feedstock.

Over the past several years, the company has become more vocal on the need to tailor regulatory mechanisms for different sectors of the economy. For instance, while DuPont supports market mechanisms such as emissions trading or tax incentives as an effective way to distribute capital efficiently, it believes it is necessary to delineate between the manufacturing, buildings and transportation sectors due to differing price elasticity and responses to price signals in terms of GHG reductions. Otherwise, one sector (such as transportation) might bid carbon prices to a level high enough to adversely impact another sector (such as manufacturing) while not making the needed GHG reductions. It is critical to balance the need for reductions across all sectors with awareness that economic ramifications are unequal across sectors. And McFarland is quick to add, "You've got to get consumer emissions under control if you ever want to get anywhere."

Challenges Ahead

DuPont has a history of energy efficiency and climate change related action that, like its overall age as a company, is much longer than most of its peers. This puts the company in a unique position. With 15 years of experience tracking GHGs and 12 years of experience in implementing emission reduction goals, the company has achieved a great deal of success in reducing its own GHG footprint. Being further along the learning curve than most allows DuPont the ability to see the next hurdle much more clearly. DuPont must successfully transition climate change and energy efficiency from an issue of risk management to one of business opportunity.

But, in DuPont's view, it is relatively easy to set goals, measure progress, learn process improvements, and find reductions in energy use. And although work on process improvements and energy efficiency projects continue, most of the big reductions have already been realized. The real challenge lies in moving beyond reductions and identifying and evaluating business opportunities in a carbon-constrained world. "Identifying market opportunities is a different challenge from footprint reduction," says Fisher. "With footprint reduction, it's easy to clarify what you want people to do—reduce X percent of what you are emitting. Alternatively, to look at 22 businesses and envision market opportunities in a carbon-constrained world is more difficult. It starts with an analysis of what you do, looking down the value chain, understanding what your customer needs and meeting those needs. As with any type of innovation, you have to make sure that new ideas will meet customer needs and satisfy regulations." Towards that end, the organization is working on developing scorecard type tools for R&D, business leadership, and marketing to use to help meet their market-facing goals and assess the improvements that products bring to customers and final consumers.

In order to understand and take advantage of this new focus, DuPont must navigate the complexity of the climate change issue, including the science, politics, economics, and uncertainty surrounding the timing. For example, rising energy prices this past winter (2006) have raised interest in green building and energy-efficient housing, but it remains unclear if energy prices will be persistently high enough to increase demand for related products from builders. Furthermore, it is unclear how policies, ranging from Federal energy policy to local building codes, will influence the marketplace. It will be a complex issue involving both push and pull from suppliers, producers, builders, end consumers and regulators.

Sharing information internally in such a large organization also remains a challenge. Despite an extraordinary organizational structure for sharing and disseminating information, the company "is still stove-piped," says Dawn Rittenhouse, Director of Sustainable Development. (This is a problem that executives feel is not unique to climate change and applies to the company in general.) Having widely distributed decision making contributes to the risk of business units acting in a bubble. The danger, especially regarding climate change and energy efficiency issues, which can be seen as add-on issues, is that technical expertise and success stories make it up, but not across, the organizational hierarchy. With the company being so diverse and products involved in almost every value chain, it can be difficult to make sure that all opportunities are identified and pursued across all of the DuPont businesses.

Another challenge is streamlining and fine-tuning their emissions measuring and tracking system, which many consider labor intensive. Energy related emissions are calculated based on fuel consumption according to the WRI/WBCSD GHG Protocol and fuel-specific measures. The current system requires input of data from direct metering of gas, invoices for other fuel purchases, reconciliation to inventories, and the application of emissions factors for a variety of fuels to calculate emissions. Process emissions are reported separately and indirect emissions must be calculated based upon localized information. All of this information is collected once per year in a corporate database. Despite having tracked emissions since 1991, DeRuyter still believes that the company's "biggest headache is in capturing and reporting data, particularly energy reporting and verification of 3rd party invoices." There is no link with the company's SAP system, which would be desirable but is currently prohibitively expensive.

In the end, the key for a science and innovation based company such as DuPont is the development of new materials that will take the company through its next transformation and into its third century. Says Fisher, "We need to understand, measure, and assess market opportunities. How do you know and communicate which products will be successful in a GHG-constrained world? How should we target our research? Can we find creative ways to use renewables? Can we change societal behavior through products and technologies? The company that answers these questions successfully will be the winner."

Weaving Climate Change into the Business Case

Alcoa*

What do you do about climate change when energy comprises a major portion of the total cost to manufacture your product? That is the dilemma that faces Alcoa, a company that spent over two billion dollars on energy last year. Consequently, locating and securing reliable, low-cost energy sources has always been among the company's most pressing strategic concerns. According to Jake Siewert, Vice President, EHS, Global Communications and Public Strategy, "the biggest differentiator in primary metals is long-term energy supply; 20 to 40 plus years." With global energy prices continuing to rise and climate change assuming a more prominent role in international policy discussions, energy intensive companies such as Alcoa are facing increased scrutiny by foreign governments, as well as stiffer competition from other companies and industries for the remaining sources of abundant, inexpensive energy.

Table 11

Alcoa's Footprint (2005)

Headquarters:	Pittsburgh, PA
Revenues:	$26.2 billion
Employees:	129,000
Percentage of Emissions In Kyoto-Ratified Countries:	16 percent
Direct CO_2e Emissions*:	34.4 MMtons**
Indirect CO_2e Emissions***:	27.0 MMtons
Aggregate CO_2 e Emissions:	61.4 MMtons
Target:	25 percent below 1990 levels (achieved and maintained since 2001)
Year Target Set:	1998

* 100 percent of the direct emissions from facilities managed by Alcoa.
** Million metric tons.
*** 100 percent of the indirect emissions associated with purchased electricity from facilities managed by Alcoa based on estimates of the sources of generation used by suppliers.

But, by leveraging its strong history of environmental leadership and responsibility, Alcoa is striving to transform this potential source of vulnerability into a competitive advantage. The company has already managed to reduce its direct greenhouse gas (GHG) emissions by 25 percent below 1990 levels and is being recognized for its progress: Alcoa was named one of the Top Green Companies in the world by *Business Week* magazine[144] in 2005 and Ceres in 2006.[145] But despite such praise, the company is looking toward two key developments that could result in further dramatic reductions: the development of a new aluminum smelting process and a vigorous effort on recycling and automobile light-weighting. When you combine what Alcoa has accomplished with the potential that lies ahead, Alcoa "is in a unique position and one that is very positive, given the attributes of the products we make," says Randy Overbey, President of Primary Metals Development.

Company Profile

Headquartered in Pittsburgh, Pennsylvania, Alcoa is the world's leading producer of primary aluminum, fabricated aluminum, and alumina. The company employs approximately 129,000 "Alcoans" (a term that employees use to refer to themselves) in 43 countries. The company earned revenues of $26.2 billion in 2005 by producing approximately 11 percent of primary aluminum in the world. Customer segments include aerospace;

* We would like to thank Pat Atkins, Ken Martchek, Richard Notte, Randy Overbey, Jake Siewert, and Vince Van Son for their contributions to this case study.

automotive; packaging, building and construction; and commercial transportation. Alcoa also produces and markets consumer brands including Reynolds Wrap®, Alcoa® wheels, and Baco® household wraps. Among its other businesses are vinyl siding, closures, fastening systems, precision castings, and electrical distribution systems for cars and trucks.

Because energy is so critical to Alcoa, the company generates approximately 25 percent of its own electricity needs. Overall, its energy supply portfolio consists of hydropower (35 percent), coal (36 percent), natural gas (18 percent), and oil (9 percent). Total direct GHG emissions from company managed facilities in 2005 were approximately 34.4 million metric tons of CO_2 equivalents (CO_2e), coming primarily from its smelting operations, power generation facilities, and refineries. In 1998, Alcoa set a target to reduce its direct GHG emissions from managed facilities 25 percent below 1990 levels, an ambitious goal when compared to other *Fortune* 500 companies actively pursuing GHG reduction strategies. Much like DuPont, Alcoa has a history of setting and attaining far-reaching targets, particularly in the environmental arena. The company achieved its GHG reduction goal in 2001 and has maintained that level ever since.

Climate Change Program Implementation

When asked about the impetus for its climate change strategy, Ken Martchek, Manager of Life Cycle and Environmental Sustainability, states that: "Sustainability is a primary driver since Alcoa defines sustainability as financial success, environmental excellence, and social responsibility. Our climate strategy is an essential part of our sustainability efforts given Alcoa's level of energy consumption particularly in an increasingly carbon-constrained world." But it is even more than that, according to Pat Atkins, Alcoa's Director of Energy Innovation. Reducing environmental impacts is smart business too. "Why wait for irreparable harm from climate change or policy requirements to make strategic and operational changes if the business case is already there? Alcoa is vulnerable because of our high energy demands and our need to grow to supply the market demands for our products. If we become part of the solution rather than part of the problem, we have a much better chance of continuing to contribute in the future. Many businesses tend to focus on the next quarter or next year, not their fourth century. Alcoa has operated in three consecutive centuries so far, and if we don't focus on climate change, we may not make it to our fourth century. Our products need to be sustainable in the broadest sense."

This attention to the long term goes to the top of the organization, an aspect not lost on those responsible for managing Alcoa's climate-related strategies. "On a scale of one to ten, senior level support is an eleven," says Atkins. "Climate change is generally not chosen as a priority unless it is supported by those at the top." While a systematic focus on energy efficiency has enabled Alcoa to reduce the amount of energy required to refine bauxite into alumina, reduce alumina into aluminum in its smelters, and fabricate aluminum into value-adding products, the primary focus of Alcoa's GHG reduction efforts thus far rests in reducing perflourocarbon (PFC) emissions through anode effects and increasing the use of recycled materials.

The Anode Effect. In 1992, then CEO Paul O'Neill, an industrial engineer turned economist (and former Treasury Secretary), asked Alcoa's engineers why the company did not eliminate the anode effect from operations (see "Anode Effect: An Overview" on this page). Believing that stable operations reduced waste, O'Neill challenged the company's engineers to eliminate the need for the anode effect by devising an alternative method for managing the aluminum smelting system. The engineers responded with skepticism, claiming the solution would be cost prohibitive, requiring thousands of new, more accurate alumina feeders, as well as better algorithms in the company's computer programs. And after all, scheduling anode effects as part of the process control scheme was the way that aluminum smelters had been run for many years. Undaunted, O'Neill continued to challenge engineers to minimize the number of anode effects, and after numerous iterations engineers discovered that new feeders were not always needed. Instead, they found that what was needed were new advanced cell control algorithms to manage the feed of alumina into the cell without having anode effects. At the same time, Alcoa signed a voluntary agreement with the EPA to reduce anode effects. With every iteration of the algorithm, control engineers noticed both a reduction of anode effects and an improvement in cell efficiency and alumina quality.

Anode Effect: An Overview

The aluminum smelting process is a highly energy intensive electrolytic reduction process used to break the atomic bond between oxygen and aluminum in alumina (aluminum oxide, Al_2O_3). The smelting process uses consumable carbon anodes to reduce the alumina creating aluminum and CO_2.

A critical aspect of the aluminum manufacturing process is maintaining the proper concentration of alumina in the electrolytic bath solution. If the alumina concentration is too high, undissolved alumina falls to the bottom of the cell causing inefficiencies and potential damage to the cell lining. If the alumina concentration is too low, the electrical current starts to break down other chemical components in the bath (namely aluminum fluoride) necessary to continue making aluminum. This reaction creates the perflourocarbon (PFC) gases CF_4 and C_2F_6 which form beneath the anode and increase the cell resistance. When the increasing resistance causes the cell voltage to exceed a threshold, the cell is said to be on "anode effect." The anode effect is not extinguished until the alumina concentration has been increased and the voltage is reduced. Anode effects have three primary drawbacks: they disrupt the stability of the continuous electrolytic process, consume excess energy, and create PFC emissions.

In the past, the level of alumina concentration in the process was routinely determined by purposefully scheduling anode effects by underfeeding alumina. This practice had provided an easy and reliable means of determining the amount of alumina in solution. The anode effect would give the plant manager an exact point of reference as to the amount of alumina in solution and helps avoid the risks and consequences of over-feeding alumina into the cell.

Alcoa no longer schedules anode effects. Although they still occur periodically, the company has reduced the anode effect frequency in its best plants from approximately one or more anode effects per cell per day to one anode effect per cell every 10 to 30 days. This reduction in frequency, coupled with reductions in anode effect duration, has reduced PFC emissions by over 75 percent since 1990. To continue to improve performance, Alcoa has company and plant-specific goals for minimizing the frequency and duration of anode effects. At some locations, a portion of each employee's annual incentive payment is tied to anode effect performance.

According to Overbey, even though Alcoa was considering the anode effect before the arrival of O'Neill, it was his leadership that made it clear that the company was ready to move beyond the climate change debate and take real action. The company needed someone to ask the right questions, help the employees overcome some in-house biases, and think about operations from a different perspective. O'Neill was that person.

Today, Alcoa is working to develop a new smelting technology based on an inert anode, which would eliminate consumable carbon anodes and related CO_2 emissions from the aluminum smelting process, and also eliminate all PFC emissions.

Recycling Initiatives. In a 2002 speech to the U.S. Aluminum Association, John Pizzey, then Group VP for Primary Products, argued that it was his fundamental responsibility to effectively manage climate change, energy reduction, and water quality issues. He then pledged that 50 percent of Alcoa's products, other than raw ingot sold to others, would come from recycled aluminum by 2020. According to Overbey, "recycling is not only the right environmental choice; it can be the right economic choice for Alcoa." Considering that aluminum produced from recycled materials requires only five percent of the energy needed to make primary aluminum and that energy prices will likely continue to rise, increasing recycling rates is among the more significant long-term strategic opportunities for the company. Almost 70 percent of the aluminum ever produced is still in use today, totaling approximately 480 million metric tons. The amount of aluminum recycled in 2004 equaled the total amount of primary aluminum produced in 1974.

To meet its target for higher recycled content, Alcoa will have to overcome some of the challenges that have traditionally undermined recycling initiatives. In addition to resolving some metallurgy issues associated with recycling, Alcoa will need to devise innovative strategies for collecting large quantities of metal and ensuring that it satisfies the company's quality standards. Further, it will have to engage with external groups to increase aluminum can recycling rates, which in the United States have declined from well over 60 percent to 50 percent in the last few years. The company is currently reevaluating how to engage these customers by focusing on the long-term financial benefits offered by elevated recycling rates.

Organizational Integration

Alcoa relies on three dedicated teams to further its climate change and energy efficiency goals: Corporate Climate Change Strategy Team, Greenhouse Gas Network and Energy Efficiency Network. These teams complement each other under the umbrella of Alcoa's values and drive to share best practices across the company. And, although the company has a long standing culture of technology and best practice transfer, employee engagement is crucial. According to Vince Van Son, Manager of Environmental Finance and Business Development, "Our people link our systems and our success. The best technology only gets you so far. Employees will devise innovative ways to achieve clearly stated goals when they understand the linkage with the company's vision and values." Similarly, according to Atkins, Alcoa's managers are becoming more aware of the importance of the company's strategy because they understand how climate change impacts their respective business units. They realize that, "if you want to build a new plant, having Alcoa's reputation helps."

In 1997, Alcoa launched the Corporate Climate Change Strategy Team. Traditionally directed by some of Alcoa's top-level executives, the team is comprised of eleven diverse members, including professional representation from operations, government affairs, technology, communications and finance and geographic representation from the

United States, Canada, Australia, Europe, and Brazil. The team is responsible for evaluating the impacts of climate change on Alcoa's business interests and disseminating the company's goals and progress to internal and external audiences. According to Overbey, the current director, the secret to the success of the team is its multi-functional membership. "The members may not always agree with each other, but having such diverse representation increases the robustness of our results." The team meets face-to-face at least twice each year and conducts conference calls between meetings.

To build on the success of reaching its goal of reducing GHG emissions 25 percent below 1990 levels, Alcoa launched the Greenhouse Gas Network in 2004 to help further reduce GHGs among locations involved in power generation, refining, and smelting—which collectively account for approximately 90 percent of Alcoa's total emissions. The network works with global process technology teams and various regional GHG teams across the world to coordinate and share information and best practices.

One of the most important projects of the Greenhouse Gas Network is the recent launch of an internal web-based GHG information system. Alcoa has systematically collected GHG data for all operations worldwide since 1998 through its environmental data system. This system makes it easier for locations to monitor their performance through time and compare it relative to internal and external benchmarks. By increasing overall transparency, the information system provides underperforming plants with a stronger incentive to improve efficiency and to lower GHG emissions. Centralizing GHG emissions accounting also promotes consistency with protocols and enables locations to focus resources on making reductions. Alcoa also relies on the system to facilitate global networking among the participants and help stimulate sharing of best practices.

The new GHG information system currently includes detailed process and energy consumption information for 41 facilities worldwide, including four power generation facilities, nine alumina refineries, and 26 smelters. The system uses the methodology of the E.U. ETS to calculate emissions and sweeps databases each evening to pull process and production data directly. Designated individuals at each plant are responsible for manually entering energy consumption data on a monthly basis. Reminders to update monthly energy data are issued automatically to help ensure a comprehensive overview of the performance of all facilities is available as soon as possible after the end of each month.

One example of leadership and sponsorship for GHG emission reduction is a global PFC reduction "challenge." Each plant has been challenged with closing the gap between its 2004 average anode effect (and thus PFC emissions) performance and its best monthly performance on record. Each month, a global scorecard is published comparing smelters to themselves and others in terms of CO_2e emissions, CO_2e emissions per ton of aluminum, and anode effect performance. As the scorecard is a highly visible way to track the leaders and laggards, it fosters healthy competition on GHG reduction progress as plant managers strive to have their facilities be leaders. But, the company also encourages cooperation and cohesion by mandating that each facility disclose barriers to meeting its targets, as well as the actions they are taking to overcome them and reach their targets. By emphasizing transparency and the sharing of best practices, the company ensures the focus on meeting targets is sustained. Some of the plants making the greatest amount of improvement to date have done so in part by sending employees to smelters that have lower emissions to learn first-hand of the latest changes they have made to cell operating, maintenance and control practices.

Similar to Shell's Energise program, Alcoa launched the Energy Efficiency Network (EEN) in 2002. The EEN consists of more than 450 Alcoans worldwide. Individual teams are comprised of high-level Alcoa employees and various external experts. These teams have three roles. They are invited by operating locations to conduct energy efficiency assessments that confirm and help solidify the business case for possible improvement. They also identify, document and distribute globally, any strong energy practices observed at the plant locations. Finally, they provide technical support and access to further resources as needed. As of mid-2005, assessments had been completed at more than 50 plants. To increase ownership of assessment results, plants participate in reviewing initial recommendations and reach agreement on potential savings before a final report and action plan is issued. Plants have confirmed nearly $80 million in annual savings potential and captured sustainable annual savings exceeding $20 million.

The overall goal of the EEN is to help Alcoans understand and value the long-term benefits of the company's energy efficiency and climate change strategy. According to Vice President of Energy Services Richard Notte, who manages a global portfolio of energy for Alcoa and the EEN, "The challenge is to provide a business case which influences us to make a shift on how we think about energy. The Network, including all its members globally, encourages us to consider energy as a manageable expense and also to consider life-cycle costs. As that shift occurs, we build into our business concept that production can be met using less energy per unit of output without sacrificing quality or production. Also, it encourages us to know and consider the cost of energy flowing through the equipment when making maintenance and end of life equipment replacement decisions. This shift in thinking provides significant financial and environmental paybacks." Although the precise cost of setting up the EEN is difficult to quantify, Alcoa estimates are as high as $500,000 after accounting for travel, human capital, and use of internal resources. Notte believes that the vast majority of companies will not require the same level of sophistication as Alcoa. "Our system is as complicated as anyone is going to get."

Projects requiring capital investment are pursued based on their financial return and the fit with the other local needs and strategic interests. The availability of capital and the threshold financial return or internal rate of return (IRR) required therefore depends on the business situation of the individual location. The company has traditionally not pursued such projects unless they have had a payback of one year or less. However, as the program has matured, provided real returns and demonstrated its potential, Alcoa is moving beyond the "low hanging fruit" and investing in projects with longer payback periods. Plants have been asked to keep track of all energy efficiency projects as they can become more attractive with time (as energy prices rise). Within the Primary Metals division energy efficiency projects with an IRR as low as 20 percent can be considered even if needed funds might not be allocated as part of a given plant's annual capital budget. According to Van Son, the identification and tracking process is critical: "The most important step is to get all opportunities systematically on the radar screen. Just as every piece of fruit ripens at a different time, not all projects should be pursued immediately. The process starts with quality information."

Alcoa has also taken significant steps to extend the reach of its climate change strategy beyond operations and into the personal lives of Alcoans in an effort to help broaden engagement in the issue. For example, following on the heels of its successful One Million Tree program the company launched an even more ambitious Ten Million Tree program on Earth Day in 2003 to help increase employee awareness about climate change, carbon sequestration, and the importance of reducing GHGs. To reach the goal of planting 10 million trees by 2020, each participating Alcoa location purchases the trees from a supplier of choice and distributes them to employees. Alcoans are then encouraged to plant the trees in their communities and on Alcoa property. Through 2005 the number of trees planted via these internal programs is estimated at 1.5 million. The company also plants millions of trees each year as a part of its mine reclamation projects around the world.

Alcoa has also encouraged its employees to participate in local and regional programs such as Smart Trips[146] to encourage use of public transportation and car pooling and the one-ton challenge launched by the Canadian Government in 2003. The one-ton challenge enables individuals to measure their GHG footprint and pledge to pursue those actions they can take to reduce their personal emissions by one ton per year.[147]

External Outreach

As with other companies in this survey, Alcoa's climate-related strategy reflects, in part, the insights it gains from its external outreach. To accomplish this, Alcoa has formed partnerships with various environmental non-governmental organizations (NGOs). Although the company acknowledges that such partnerships provide the company with credibility and third-party verification, it emphasizes that these relationships are much more than just stamps of approval. According to Siewert, "We know we're not the expert on these issues; we need help. Our people broaden their view of sustainability by interacting with others who think more broadly, with the people who help manage the growth process more effectively. When we think too narrowly, we get in trouble because the rest of the world doesn't think that way." Martchek believes these partnerships also provide the company with more leverage to participate in the process of shaping climate change policy. "Working closely with organizations like the World Resources Institute and Pew Center on Global Climate Change provides us with some insights about what the future may look like."

Moving beyond environmental NGOs, Alcoa has worked with several external groups to further its goals of increased recycling. The company is a member of the Curbside Value Partnership (CVP). CVP is an outgrowth of the Aluminum Can Council, a trade organization comprised of companies that make aluminum can sheet and aluminum cans. CVP joins with large and small communities across the United States, and their material recovery facilities, to increase education and promotion of recycling of a variety of valuable materials through existing curbside collection channels. CVP assists communities with participant education and promotion, data collection and interpretation and understanding the value proposition of recycling, especially aluminum can recycling. While proven to increase recycling rates, deposit legislation has traditionally been opposed by some of Alcoa's largest business customers. Alcoa and many of its customers favor a more comprehensive approach to recycling, such as that advocated through the Curbside Value Partnership. And finally, aluminum can sheet makers continue a partnership with Habitat for Humanity, which channels money earned from recycled cans into materials for homes constructed by Habitat.

Alcoa routinely seeks the input of its key investors. Since 2003, Alcoa has convened its top five to fifteen investors during one to two visits to key facilities each year. During these visits the company has frank discussions about its corporate governance and sustainability initiatives. These events are an integral part of its communications and investor relations strategy. In addition, Alcoa's annual sustainability report is used by analysts and other interested stakeholders and documents how the company mitigates risk by reducing its footprint.

A final prong in Alcoa's outreach is directed at the aluminum industry; a highly consolidated industry that offers a potential opportunity. As Siewert explains, "At any time, the aluminum industry can easily get 75 percent of world capacity at one table. This is not true of other industries." But despite such high industry consolidation, the industry lacks a consistent strategy or approach to addressing climate change or energy issues. Therefore, Alcoa recognizes a value both in making great strides in emissions reductions and encouraging others to follow. Mindful of competition from cheaper, less energy-intensive metals, Alcoa believes it is in its own economic interest to raise the reputation and standards of the entire aluminum industry, particularly in places like Europe. And Alcoa's international competitors are beginning to respond to the challenge by improving efficiency and reducing emissions. To increase access to certain financial markets, competitors from Russia, China, and the Middle East are increasing transparency of operations by publishing sustainability reports.

Policy Perspectives

In general, Alcoa supports cap-and-trade systems where regulatory limits are imposed if all gases are included. Alcoa currently empowers local management to determine the company's official position within each country. And elements of these positions can vary based on local circumstances.

Of greatest concern to Alcoa is climate change legislation that does not recognize companies for taking early action. Alcoa seeks the use of a 1990 baseline for determining allocations. According to Siewert, "Although I can't imagine anything coming out of Washington that would be too strict for us, the worst case scenario is not getting credit for what we've already done." It is for this reason that Alcoa is concerned with the U.S. Department of Energy's 1605(b) program. Alcoa believes the recent DOE decision to disallow any reduction before 2003 not only discourages companies from taking early action, but potentially encourages increases in the short term.

To prod federal action, Alcoa testified on behalf of the McCain-Lieberman Climate Stewardship Act in 2003. The company feels strongly that there must be a global standard and uniform playing field for all companies. According to Siewert, "We need to know that what happens will happen to everybody." In 2005, Alcoa called for a comprehensive national registry and mandatory emissions reporting as its internal successes have shown measurement and reporting are a fundamental part of attaining any target.

Unlike Whirlpool, which seeks to retain credits for the improvements in energy consumption its products may offer, Alcoa does not lobby for gaining credits for emission reductions by users of its products. Since Alcoa mostly produces semi-fabricated products and not final products as Whirlpool does, the company is satisfied with increased sales if GHG reduction goals increase the market for their product. Alcoa believes that the high performance-to-mass ratio of aluminum products will become increasingly attractive to its transportation customers (such as autos, trucks, rail cars, and planes) in a more carbon and energy constrained world. This reinforces an already strong business case for aluminum, and market pull for its rolled sheet, extrusions, cast components, forged wheels, and other related products. While airplanes are comprised of 90 percent aluminum and titanium, the composition for automobiles is only about 10 percent. Reducing a vehicle's weight by 10 percent typically yields a seven percent reduction in GHG emissions. Based on current growth rates, Alcoa projects that light-weighting coupled with increased recycling by the global aluminum industry has the potential to offset all industry direct and indirect emissions by 2017. Lighter cars and resulting improvements in fuel economy and lower emissions can potentially save 400 million metric tons of CO_2e. To increase the demand for aluminum, Alcoa supports both GHG reduction standards and federal Corporate Average Fuel Economy (CAFE) size-based standards for fuel economy as size and intelligent design have shown to help improve passenger safety and fuel economy (and subsequently reduce GHG emissions).

Challenges Ahead

Of all accomplishments in the area of GHG reductions, Alcoans acknowledge that the development of its web-based systems for measuring and tracking emissions reductions is a major step forward in both achieving its goals and making all locations aware of their carbon footprint.

And the company is pleased it has leveraged its efforts on climate change and other sustainability issues, leading to reputation and strategic benefits. For example, the company was invited by the Icelandic government to build a smelting facility in their country; a country with an extremely low GHG electricity profile and low energy prices. Alcoa's growth in Iceland is a direct reflection of its preference to use renewable energy resources (hydroelectric power) to achieve the lowest total GHG intensity per ton of aluminum possible. When Alcoa's new smelter begins operations in 2007 it will become one of the lowest GHG intensity smelters in the world. Another example of gaining from its efforts is being recognized by Innovest (along with Toyota and BP) as the world's top three most sustainable companies. The rankings were based upon how effectively companies have managed strategic profit opportunities by recognizing new environmental and social markets. Shortly after the rankings were released, Toyota approached Alcoa to discuss potential partnerships and synergies between the companies—again, a strategic aspect of the company's future plans.

Looking forward, Alcoa seeks to make increasing progress into the light-weighting of vehicles. And the market looks bright. Over the past decade, the demand for aluminum has increased at a compound annual growth rate of 4.3 percent, and the aluminum is becoming the second most used material (overtaking iron) in ground transportation vehicles after steel. For example, Alcoa developed "Dura Bright" commercial truck wheels that are lower mass than conventional wheels and don't require polish or scrubbing. These wheels have high strength to mass ratio, are visually attractive, corrosion-resistant, and require no maintenance beyond spraying with soap and water. In February 2005, Alcoa announced that Hyundai Motor Manufacturing America (HMMA) will use an Alcoa cast aluminum rear upper control arm for the Korean automaker's all-new 2007 Santa Fe crossover vehicle; the first Alcoa component to be used by Hyundai Automotive.

But, Alcoa is still working to improve its GHG related-strategies. Despite recent initiatives to engage and educate its employees, some managers believe the company would have benefited from launching such programs much earlier. Atkins admits that the company would be even further ahead if we'd "done this in year two, instead of year ten. It takes time to educate 130,000 people." And looking forward, Overbey worries about the fact that the company's global reach truly requires a global answer to the GHG issue. This highlights one important challenge for the future. Political and regulatory uncertainty via the absence of a uniform global climate change policy creates an uneven playing field with regard to its global operations. Alcoa believes such uncertainty coupled with high energy prices provide a disincentive for companies to set up new operations in many developed countries.

But despite such challenges, Alcoa sees climate change as a major differentiating factor in the future. According to Overbey, "Whatever enterprise you represent, you must ask 'How can I be part of the solution?'" Adds Atkins, "What would the best company in the world do? We are citizens of the world and we must act responsibly." With this as its starting point, Alcoa continues to move forward through leadership and action to be part of the solution—and sees benefits in reinforcing its reputation for doing so.

Maintaining a Seat at the Table

The Shell Group*

Royal Dutch Shell, like all major oil producers, finds itself at the heart of the debate over climate change. In 2005, Shell's own operations emitted 105 million metric tons of CO_2 equivalents (CO_2e). The downstream combustion of the fossil fuels it produces emits another 763[148] million metric tons. Together these emissions account for some 3.6 percent of global fossil-fuel CO_2 emissions in any year—a total greater than that of the entire United Kingdom. But rather than sit on the sidelines and wait for carbon constraints to alter the company's business environment, Shell took an early position on the issue and engaged in actions that began to manage its carbon footprint. These actions have earned the company credibility and a powerful voice within policy, advocacy and market circles. And this voice grants the company a measure of control over its future business environment. In the words of David Hone, Group Climate Change Advisor, "To validly have a seat at the table, you have to bring experience. You cannot just take a seat because you are interested."

Table 12

Shell's Footprint (2005)

Headquarters:	The Hague, NL
Revenues:	$307 billion
Employees:	112,000
Percentage of Emissions In Kyoto Ratified Countries:	~30 percent
Direct CO_2e Emissions:	105 MMtons*
Target**:	10 percent below 1990 by 2002 5 percent below 1990 by 2010
Indirect CO_2e Emissions***:	763 MMtons
Aggregate CO_2e Emissions:	868 MMtons
Year Target Set:	1998 Revisited in 2002 Recast in 2005/6

* Million metric tons.
** Direct emissions reductions only.
*** Measured as emissions from product use in 2002.

In order to maintain that seat, the company must continue to develop the breadth and depth of its climate change program. The company now finds itself facing the challenge of integrating what had historically been treated as two separate tracks—energy strategy and climate change strategy. Shell is seeking ways to merge the two tracks into one synergistic approach that helps them explore new business opportunities. This harmonization of strategies must also coordinate the activities of units stretched around the globe, ensuring information sharing that takes advantage of Shell's wide and varied technical expertise.

Company Profile

Royal Dutch Shell plc operates in over 140 countries and employs 112,000 people. Shell is headquartered in The Hague, Netherlands and organized into six operational units: Downstream (oil refining, marketing and chemicals); Exploration and Production; Gas and Power; Renewables (including hydrogen and carbon management); Trading; and Shell Global Solutions (technology services). The executive directors of the first three (and most important) business units also sit on an executive committee, the head of which is the CEO of Shell. The primary developer of the Shell climate change strategy has historically been Corporate Affairs, which reports to

* We would like to thank David Hone for his contributions to this case study.

the CEO. More recently, to reflect the growing importance of climate change as a strategic issue, the company has developed a new CO_2 Unit. In addition, all parts of Shell coordinate on the issue through a "CO_2 Forum."

The culture of the company centers on technology and trading, but there is also a strong sense of corporate social responsibility. In the words of Hone, "Concern [for climate change] goes quite deep. There is expectation among employees that the company is in a sustainable-development mindset. They see it as a positive thing, although it may vary by region. Employees expect Shell to uphold a high standard on progressive issues about how a company is supposed to behave."

Climate Change Program Implementation

Shell has been watching climate change since the early 1990's through its Issues Management team, a group within Corporate Affairs that monitors issues that may impact the business units. In 1998, Jeroen van der Veer, then a group managing director (and now CEO), championed a more formal study of climate change and its potential impact on Shell businesses globally. This study came after the 1997 signing of the Kyoto Protocol and at a time when the company was feeling bruised over its 1996 fight with Greenpeace over the disposal of the Brent Spar oil platform. A cross-functional team that spanned the company was put together and made the business case for implementation of a greenhouse gas (GHG) management strategy. This study raised the bar for climate action and, as a result, created resistors—"There's always a challenge to what you create," Hone says, "but building a strong business case is key to overcoming this resistance." The business case revolved around the trio of ideas that the company would eventually face a real price for carbon, that a leadership position on climate change would be a business opportunity in terms of building brand and reputation, and that a seat at the table with the governments that would set the rules was important for the company's future. Out of this initiative emerged the goal of "Securing Shell's future by seizing opportunities that arise from the climate change issue." Achieving this goal has historically followed two tracks.

The first track, energy strategy, considers the Shell energy portfolio. Planning for energy diversification is led in part by the company's well-established long-range planning tools like the Shell Scenarios (see "Shell Scenarios" on page 115). Like Alcoa, Shell has long thought in time horizons of half a century or more. And climate change requires a similarly long-term focus. "You can't look at this issue in a five-year time frame, it's almost meaningless," says Hone. "But you can look at it in a 25-year time frame—there's the scope for it to be different."

The second track, climate change strategy, focuses on managing the carbon footprint of Shell, sharing experience and validating the company's position on climate change with governments, the NGO community and the general public. The goals of this track are to build capacity for action within the company and to participate in policy development. Recognizing that carbon would have value in the future, the company began by first, taking inventory of GHG emissions, second, developing a proficiency in carbon trading and third, integrating carbon values into financial decision-making. The logic is that there will be a business benefit to both developing the experience of operating in a carbon market and working with governments to help develop those markets.

Following the 1998 study, Shell set a long-term goal of matching the Kyoto standards of a five percent reduction in GHG emissions by 2010. The first target within that goal was a 10 percent GHG reduction by 2002. This was the first hard target for Shell and it was achieved through the elimination of associated gas venting at oil production units and the reduction of associated gas disposal by continuous flaring. The second hard target (remaining five percent below 1990 emissions through the year 2010) was a more difficult sell than the first. To address internal sentiments that the company had done enough and that further public action was unwise, with the company's various business units as well as discussions with senior leaders were arranged. The workshops considered various target-setting and implementation options for the units themselves. The greatest resistance to the idea came from business units with significant growth opportunities in their forward plans. As such, a point of significant debate centered on whether to measure emission reductions targets through an absolute (for example, MMtonsCO_2e) or indexed approach (for example, MMtonsCO_2e per unit of revenue or product). Shell decided that setting one universal standard for such a large company would be impractical, as it overlooked the company's very size and the challenge that size creates. The company chose a blend of these two approaches. Individual business units would use indexed or energy efficiency measures while the Group as a whole faced an absolute target.

Shell Scenarios

Shell uses scenario planning as a strategic framework for thinking through challenges and identifying risks and opportunities. The most recent (2005) edition of Shell's scenarios, *Shell Global Scenarios to 2025,* articulates a vision of how worldwide forces might shape markets over the next two decades. The development of scenarios provides the company with a toolkit to assess risks, make investment decisions, develop a common strategic language for leadership teams, and engage in key public policy matters. Like the other Shell Scenarios, the 2005 edition uses alternative parallel story lines to explore how politics, economics and technology relate to its energy and energy services business. Shell uses story lines because stories are how humans understand the world, and stories allow for multiple levels of understanding while still giving emotional and intellectual impact. This time, for example, the three stories are: Flags, a "dogmatic, follow-me world"; Open Doors, a "pragmatic, know-me world"; Low Trust Globalization, a "legalistic, prove-it-to-me world". Through the lens of these three stories, Shell looks at issues from the U.S.-E.U.-China power balance to climate change and biodiversity. On carbon, all three stories come to the same conclusion: the world (and companies) will face a price for carbon. Practically speaking, for Shell's strategy this means focusing on increased natural gas production (especially liquefied natural gas—LNG), wind, solar, bio-fuels, coal gasification and experimentation with hydrogen delivery systems. But Shell emphasizes that it is still working to make its core business—fossil fuels—succeed in a carbon-constrained world.

To reach its first target, Shell looked first at the lowest-hanging fruit, achieving a sizable portion of its pre-2002 emissions reductions by ending the venting of associated gas (methane) from its exploration and production facilities and, most significantly, from its Nigerian operations. As the company heads toward its 2010 target, the emphasis has shifted to ending the flaring of the same gas. The company devotes energy and resources into capturing these gases and either pumping them back underground or feeding them into nearby facilities for small power stations. When the economics are right, these gases can also be converted into LNG, a major growth area for the company. Through these actions, Shell hopes to reduce its CO_2 emissions by a further 13 MMtons (from 2005), but recognizes that this reduction makes room for future growth, such as the expansion of its oil sands facilities in Canada. Shell had a global goal of ending all but small-scale continuous flaring of associated gas by 2008 but has said that it will miss this deadline in Nigeria, where the government has set elimination of flaring as a country-wide goal.[149]

The Group wanted to involve all operations in its efforts to meet the second GHG target and wanted to shift attention away from a sole focus on gas flaring. So, it sought further involvement and further reductions through individualized attention to energy use at local units. To spur reductions, Shell has set 2002 to 2007 energy efficiency targets in the refining and chemicals operations at five and eight percent improvements, respectively.

For this effort, the company also engaged its internal consulting arm, Shell Global Solutions (SGSi). SGSi consultants have helped develop many of the Group's technical solutions while also offering its consulting expertise to external clients. The consultants can be called in for projects as large as the design of refineries or as small as individual unit efficiencies. One of the SGSi programs, Energise, works specifically on energy efficiency strategies. At the request of unit managers—typically at refineries—Energise deploys teams to evaluate possible efficiency improvements. The work of these teams is similar to Alcoa's Energy Efficiency Team, which recommends operational, equipment and behavioral changes. Site management decides whether and how to implement the recommendations. Energise personnel are drawn from all areas of Shell, giving a broad range of technical expertise.

To gain access to available capital, energy efficiency and GHG emissions reduction projects must meet the same internal hurdle rate as other investments. However, the company uses internal shadow prices for carbon in evaluating its investments that then give such projects additional impetus. Shell currently uses three different (proprietary) carbon prices for valuing climate change in its investment decisions; one for the E.U., a second for other developed countries and a third for the developing world. Mandatory carbon regimes such as the Kyoto Protocol have helped to drive these internal pricing models and have made GHG and energy efficiency projects more attractive on a bottom-line basis since GHG emissions now have a real price in an external market.

By way of illustration Hone explains how the value of carbon can be a significant driver in energy efficiency decisions. One barrel of oil produces about 0.36 metric tons of CO_2. At current (early 2006) crude prices of around $60/bbl, an E.U. Emissions Trading Scheme (E.U. ETS) CO_2 price of 25 Euros is like adding a further $11/bbl to the price of oil, which makes an energy saving project even more compelling. The company uses long-term premise values for both oil and carbon when valuing internal efficiency projects (the actual numbers used by Shell are confidential and change with the market).

But to realize the full benefits of carbon shadow pricing and monetize the cost of carbon, emissions trading has become an important prong of Shell's strategy. "It is an enabler of energy efficiency projects," states Hone. For that reason, the company was one of the early innovators in both internal and external GHG emissions allowance trading. These experiences are a good example of how the climate change issue started at the periphery of the company and moved to the core of its operations. Carbon trading began as an issue for the Health, Safety & Environment (HSE) group within Corporate Affairs with the creation of a company-wide internal trading system (ended in 2002), and then for Shell Trading with creation of a CO_2 trading desk at the end of 2001. The new trading desk allowed Shell to participate in both the Danish and U.K. emissions trading schemes, which ran prior to the E.U. ETS, hence gaining valuable experience. Shell made the first swap between the Danish and UK systems in 2002 and, while the market did not formally open until 2005, Shell made the first actual market trade in E.U. Allowances in 2003. By moving from HSE to Shell Trading, "GHG is becoming more and more internalized" according to Hone.

The results of Shell's internal trading experience are mixed. They show less-than-satisfactory results on its intended outcome: gaining the greatest reductions at the lowest cost (see "Internal Trading Shows Limited Success" on this page). But the company feels that internal trading was successful in making people aware of the need to reduce GHG emissions and the use of trading mechanisms to do it. This expertise also gave Shell credibility in policy circles and meant that its views were considered in the development of the E.U. ETS that went into effect in 2005.

Internal Trading Shows Limited Success[150]

The Shell Tradable Emissions Permit System (STEPS), the company's first attempt at GHG emissions trading, had decidedly mixed results. Begun in 2000, STEPS was an internal cap-and-trade scheme designed to last three years. Units within Shell joined STEPS voluntarily and were allocated tradable emissions permits based on their past history of emissions. These units accounted for 70 percent of Shell's emissions in Kyoto Annex I countries. The goal was to reduce the emissions of these units to 2 percent below 1998 levels using declining caps on permit allocations under the trading system.

STEPS offered some benefits to the company. It gave Shell's units practical experience in both trading and calculating the cost curves for GHG abatement. The program also helped train Shell units for mandatory trading systems under the E.U. and Kyoto. While it provided these benefits, the program did not live up to expectations for several reasons:

1) The voluntary nature of the program meant there were not enough participants and not enough liquidity in the permits market. Only units that could easily reduce their emissions tended to participate—making the market price for permits artificially low.

2) Shell units in different countries could not monetize the internal GHG emissions trades because of the tax liability it would generate.

3) Midway through the scheme, some units asked for—and received—extra permits from company headquarters. This "going back to the government" created uncertainty and softness in the already illiquid market.

Beyond its internal and external trading, Shell also became actively involved in early initiatives under the Kyoto Protocol's Clean Development Mechanism (CDM). Initial success here was also limited. The company faced problems both related to the CDM structure and of their own making. In one solar project, the company determined that the cost of going through the CDM process exceeded the benefits of the carbon offsets. In an energy efficiency project in Buenos Aires, the company has been in the CDM Executive Board process for over a year (as of January 2006), leading to some frustration with the process. In addition, Hone feels that the Group "wasted" effort on early (1999-2000) internal CDM workshops but couldn't produce concrete results because of the slow start to the CDM market. Now, with the CDM market emerging and beginning to look like a success story, the company is working to reengage its businesses and capitalize on the opportunities that CDM offers. Early in 2006 Shell Trading was the recipient of the first physical forwarding of Certified Emission Reductions to an account on the United Nations Framework Convention on Climate Change (UNFCCC) Secretariat's Clean Development Mechanism Registry.

Beginning in 2005, the company found itself at a crucial crossroad as the carbon issue began to figure significantly in the Group's forward-looking strategy. An internal CO_2 Study concluded that the Group must step up its efforts on GHGs. It must find ways to integrate its energy strategy and climate change strategy tracks into one cohesive strategy that helps the company identify and capture new business opportunities as well as maintain its core fossil fuel business.

In a January, 2006 *Financial Times* editorial,[151] Shell CEO Jeroen van der Veer articulated Shell's conclusion that future production of liquid fossil fuels would increasingly depend on unconventional sources, such as oil sands, gas-to-liquids, oil shale and coal gasification. The days of "easy oil" are over. The more difficult oil is "dirtier" and the company will subsequently have to address its associated higher GHG output. Van der Veer stresses the importance of carbon sequestration—both underground and combined with other materials to make inert materials, as a technical solution. It has become clear that the energy portfolio will have a significant impact on its GHG profile. Conversely, the company's climate change strategy has created the expectation of a company able to manage GHG emissions and government action has created carbon value in the market. These two tracks must now be intertwined. The Group's future depends on it.

One important acknowledgement of this increased importance is the creation of a new CO_2 unit led by a senior executive. Graeme Sweeney, also head of Hydrogen and Renewables at Shell, has filled the post. His role will be to attend to the development of Shell's CO_2 strategy and the technologies that support it. The Group's CO_2 unit under Sweeney is viewed as a place to kick-start and foster GHG reduction technology until it is sufficiently integrated in the business units to stand on its own.

External demand for lower-carbon energy has led the group to look toward key growth product lines. The first is continued attention to "developing LNG and natural gas businesses as a very easy way to help transition to a low carbon world" since natural gas has half the carbon footprint of coal in electricity production. As part of its broader energy portfolio, Shell has a strategy of having many technological irons in the fire. "A lot of energy technologies have come and gone," Hone says, "and it's hard to predict what the next big hit will be." The company has invested over $1 billion in new technologies such as wind, solar, bio-fuels and hydrogen, and is now stepping up investment in underground sequestration and IGCC/coal gasification.

Within the last two years, there has been a growing realization that coal is going to be an integral part of the global energy mix, particularly in China and India. As gasification is a chemical conversion, an existing proficiency of the company, and has applications across a broad range of products and markets, the company sees a significant opportunity in this area. Shell's experience with gasification dates back to the 1950's when the first gasification unit was commissioned with oil as feedstock. There are now over 150 Shell Gasification Process (SGP) gasifiers licensed worldwide. The experience gained on oil gasification provided a firm theoretical and practical base for the start of the coal gasification development in 1972. In fact, the technology has been utilized in a coal gasification pilot plant in the Netherlands. The process can be used to make "syngas" which can be used to make everything from electricity to plastics and importantly liquid transport fuels or even hydrogen for transport. Further, the process could be altered (with further R&D) to accommodate feedstocks of wood chips, municipal waste or other materials that could be gasified into useable fuels.

The company's solar operations are an outgrowth of solar research that started after the energy crises of the 1970's and is now focused primarily on non-silicon based, copper indium diselenide (CIS) "thin-film" panels. Shell is also one of the ten largest wind farm owners in the world with capacity greater than 350 MW. Its wind portfolio is planned to grow at the market rate of expansion to 500 MW by 2007.

However, as advances are made, the company finds that some renewables clash with the existing business model. For example, electricity generation is not part of Shell's core business, yet wind power is fundamentally an electricity business. Similarly, Shell Solar has undergone both expansions and contractions, buying Siemens Solar in 2001 and then selling its silicon-based solar activities in 2006 to SolarWorld AG. The remaining thin-film business line has sought a partner in the form of Saint-Gobain, a company with "film-on-glass" technology expertise. And, as Hone puts it, "Can an oil company like Shell compete in a market where an electronics company like Sony or Sharp can bring a lot of R&D and manufacturing expertise to bear?"

Hydrogen production is an area where Shell is developing critical expertise and is seeking to leverage that expertise in its investments. Shell already produces 7,000 tons of hydrogen per day, mostly from natural gas, and mostly for use in refinery operations. Shell hopes to use this existing source of hydrogen in some of its early efforts to make hydrogen more widely used as a fuel. Right now, says Hone, "98 percent of homes within the E.U. are within 100 kilometers of someone's hydrogen production site." Because the existing infrastructure is already there, all that is necessary for this opportunity to realize itself is an awakened demand and continued refinement in hydrogen handling and distribution technology. Before that happens, Shell recognizes that it needs to be up-and-running and prepared to meet the demand. So, for example, the Group now operates four hydrogen filling stations—in Tokyo; Amsterdam; Washington, DC; and Reykjavik—and is planning to build one in Shanghai in partnership with Tongji University. Further stations are also planned for the United States.

Organizational Integration

To help diffuse and incentivize climate change initiatives, Shell has incorporated climate change related goals into individual business scorecards. Scorecards use a number of criteria to evaluate performance of business units and individual managers, and focus on two or three principal metrics, such as financial performance. A particular climate change initiative (e.g. preparation for the E.U. ETS by E.U. refineries) might account for five percent of a given score in a particular year—an amount Hone describes as "modest". But the measures are constantly changing, reflecting a particular year's goals. The scorecards are used for calculating bonuses more so than promotions and are revised each year to reflect new concerns.

Beyond scorecards, three other devices foster information flow and innovation: the Annual Report, the *Shell Sustainability Report* and an internal *Climate Change Newsletter.*

The *Shell Sustainability Report,* produced annually, serves three purposes: to be the company's public face, reporting its activities to the outside world; to act as an internal coordinating mechanism, giving staff and the various business units a guiding vision; and to allow those units to communicate their concerns and ideas during the process of compiling the *Report.* To develop the report, which is published each year in April, cross-business workshops are organized the preceding October to identify key issues to discuss and report on. "The goal is not simply to record accomplishments or make people feel good," says Hone. "It is meant to be self-challenging."

The *Climate Change Newsletter* is a purely internal e-mail document that reaches a community of 300 or more employees each month. Employees with an interest in climate change issues can find out about the newsletter on the Shell internal climate change website and subscribe. The newsletter discusses specific technologies, developments within the company, and external climate change information. Anyone within the company can receive the newsletter, yet subscribers tend to come from four categories: corporate, including legislative affairs personnel; technology development (including CO_2 sequestration and energy efficiency); commercial units such as trading; and business areas with GHG-focused projects such as the Canadian oil sands units.

External Outreach

The full gains from Shell's efforts at carbon management would not be realized without a concerted effort to engage with external groups. Shell directs its external relations regarding carbon management to four primary areas—trade associations; shareholders; NGOs and, most importantly, government.

First, Shell works through its trade associations to further develop action on climate change. At times, trade associations have taken positions that are not aligned with Shell's viewpoint. But Shell has typically chosen not to publicly break with such organizations (an exception being the Global Climate Coalition in 1998). The company instead focuses its efforts on practical measures on which there is consensus, like standardizing measures for reporting GHG emissions. Trade associations are not solely the domain of industry's large players. Hone stresses that trade associations are important to smaller players who he believes must stay involved in the regulatory development process.

Second, to allow itself the space to make forward-looking decisions about climate change, Shell believes it must convince shareholders of the merits of being environmentally responsible. The company does get climate change questions from investors and investor groups (such as the Carbon Disclosure Project) and climate change appears to be a rising issue. In addition, by watching shareholder resolutions at competitors such as ExxonMobil or Chevron, the company knows it is an issue it cannot ignore. At the request of his own internal investor relations department, Hone has given presentations to investor groups on climate change and energy development.

Third, Shell is working with NGOs on climate change issues. "NGOs," Hone says, "can expose the company to a range of views on how we are doing." Shell's work with Pew, for example, opens some doors for the company that wouldn't otherwise be available. "Once you go through Pew," Hone says, "it's like you've gone through a filtering process—you have additional credibility. Shell provides Pew with credibility. And likewise, Shell gets the same. There is less suspicion than if Shell went it alone."

Shell Canada has set up a Climate Change Advisory Panel, made up of representatives of NGOs (including a First-Nation, Native American representative) to address concerns over GHG emissions at the Athabasca oil sands project. Shell sees this as part of the integration of its energy and climate change strategies, acknowledging that this new fuel source will affect its carbon footprint, its public credibility, its unofficial license to operate and ultimately its ability to expand operations. Hone says there has been tension on the Panel from time to time, but calls it "healthy." For example, when the company was considering its second hard target on GHG reductions, the Panel was a good sounding board for ideas they were considering.

Policy Perspectives

Governments are the fourth, and most important, area at which Shell directs its external affairs activity. As governments act on climate change, Shell wants a seat at the table to discuss future regulation. "Particularly in emissions trading, these are the people you're doing a major trade with through the allocation process," Hone says of governments. "If you're doing a deal with somebody and they're setting the rules, then you want to have a say." And because climate change cuts across many issues ranging from the location of new LNG facilities to energy prices, Shell's government relations offices spend an increasing amount of their time on climate change and GHG issues with the most involvement in the U.K., strong involvement with the E.U. in Brussels and then moderate involvement in Washington. Overall, says Hone, "Our role is not to advocate that policy be enacted. We don't set policy. But if a government decides that policy is necessary, we will help them understand the best mechanisms to reach their goals."

Shell (and other corporate representatives) worked with the U.K. government to help set up the U.K. Emissions Trading Group to develop rules on trading in the U.K. Further, Shell has worked with the Corporate Leaders Group in the U.K. who, in conjunction with the Prince of Wales Business and Environment program, wrote a letter to the Prime Minister recommending more aggressive action on climate change.

Shell doesn't advocate voluntary reductions as a long-term strategy to reduce GHG emissions. "Government needs to get involved through a variety of mechanisms," says Hone. A balanced approach of market incentives, tax incentives, and subsidies is needed to create strong encouragement." Mandatory programs, such as the

E.U. ETS, will help ensure the playing field is level, define price and monetize the advances Shell makes in reducing its GHG emissions. Without the government pushing it, he says, the business case for GHG reductions is harder to make, "and action cannot take place without the business case." By contrast, Hone says, a business case driven by higher energy prices may not lead to lower carbon emissions, as higher prices may merely push companies to exploit heavier "unconventional" oil resources, dig for more coal or drill deeper oil and gas wells.

Challenges Ahead

In looking over its initiatives thus far on climate change, Hone sees the failure of the company's internal trading system as one useful lesson. While its failure was a surprise, he feels the company should have seen its limitations beforehand. But rather than dismissing the entire venture as lost, he sees benefits in the way it helped the company develop the expertise to become a leader in emissions trading in Europe.

Reflecting on all his company has done, Hone ponders, "When addressing climate change, the question is not just how will you manage your own GHGs, but how will you change the game? Ultimately, we'll have to get out of fossil fuels, but that is almost certainly many decades away. Maybe hydrogen is the answer. But you have to make the right change at the right time and in the right way. People will not get rid of cars and people will always want more energy. The key is both influencing the rules of the game and timing your shift to a new carbon-constrained strategy. It's knowing what the next technology for energy production is, and shifting when the market is ready to reward it. We're not going to get out of the oil business in the near term." But you have to ask, says Hone, "What is the iPod® for energy? Is it out there? You have to be on watch."

Don't Switch Tracks When the Train Is Already Moving

Whirlpool*

At the ninth meeting of the Conference of the Parties of the Kyoto Protocol in 2003, Whirlpool became the world's first appliance manufacturer to announce a greenhouse gas (GHG) reduction strategy. But unlike many other companies that have made similar pledges, Whirlpool's approach to climate change involves neither dramatic changes to its operations nor significant bottom-line costs. Its strategy is laser focused on leveraging its current core competencies, and continuing down the same path it has been on for years: bringing the most energy efficient products to the market and, in so doing, reducing GHG emissions through its consumers. In fact, given the company's on-going drive for energy efficiency, the words "climate change" are not often stated as an explicit concern among the workforce. The mantra is "energy efficiency" plain and simple. In the words of Mark Dahmer, Director of the Laundry Technology Division, "We've got a train moving on efficiency. We'd just start confusing things if we tried to throw more on the train or start a new train."

Table 13

Whirlpool's Footprint (2005)

Headquarters:	Benton Harbor, MI
Revenues:	$14.3 billion*
Employees:	65,682*
Percentage of Emissions in Kyoto Ratified Countries:	31 percent
Direct CO_2e Emissions:	0.8 MMtons**
Indirect CO_2e Emissions***:	146.5 MMtons
Aggregate CO_2e Emissions:	147.3 MMtons
Target:	Three percent below 1998 levels by 2008 while increasing sales 40 percent.
Year Target Set	2003

* After the 2006 Maytag acquisition, sales increased to $18 billion and employees increased to over 80,000.

** Million metric tons.

*** Measured as emissions from product use.

Company Profile

Based in Benton Harbor, MI, Whirlpool is the world's largest home appliance manufacturer. With annual sales of over $14 billion and nearly 50 manufacturing and research facilities worldwide, the company sells to consumers in more than 150 countries. Among its many major brand names are Whirlpool®, KitchenAid®, Brastemp®, Bauknecht®, and Consul®.[152] (After the acquisition of Maytag on March 31, 2006, sales increased to $18 billion and the Maytag®, and Amana® brands were added to the Whirlpool line. Whirlpool is now in the process of incorporating the Maytag emissions footprint into its overall greenhouse gas reduction target.) The company's broad vision is to have the company's products in "every home, everywhere."

Two aspects of Whirlpool's culture above all others drive the company's attention to addressing climate change. The first is a continual search for ever increasing energy efficiencies. This is born out of the company's historic focus on cost and quality in a low margin industry. The second is a close connection to its Midwestern roots, out of which emerges a strong belief in corporate citizenship. According to Dahmer, one of the core corporate principles is that there is "no right way to do a wrong thing." The company's Corporate Social Responsibility

* We would like to thank Tom Catania, Dick Conrad, Mark Dahmer, JB Hoyt, Bob Karwowski, Casey Tubman and Steve Willis for their contributions to this case study.

(CSR) statement expresses it simply as an aim to operate in "ways that honor ethical values and respect people, communities and the natural environment. Equal to protecting the health and safety of our employees, we consider environmental stewardship among our most important business responsibilities."[153] That aim is echoed in the statements of employees as the primary reason for addressing climate change. In the words of many, the company is just trying to "do the right thing." According to Dahmer, "the company is about providing for the country and the customer."

Climate Change Program Implementation

In 2003, the company announced a plan to decrease total GHG emissions from global manufacturing, product use and end-of-life by three percent from a 1998 baseline by 2008, while increasing sales by 40 percent over the same period. According to Whirlpool, these reductions were equivalent to the CO_2 emissions of 28 coal-fired plants or 10 million cars. On announcing the reductions, Tom Catania, Vice President of Government Relations, commented, "Whatever political solution the global community agrees to as the best mechanism for collectively addressing climate change, our company will continue its efforts to do our part, while at the same time bring unique, innovative and energy-efficient products to our customers."[154]

And customers are the key to Whirlpool's efforts to address climate change. Studies have shown that the majority of lifecycle GHG emissions from home appliances come from the use phase. Whirlpool's internal studies conclude that of the nearly 30 metric tons of CO_2e emitted over the life of an average washing machine, over 93 percent come from the use phase. Of the remaining amount, two percent come from manufacturing and five percent come from end-of-life disposal. This is corroborated by a 1992 study by the United Kingdom-based PA consulting group which also shows that over 93 percent of washer emissions come from use.[155]

The concentration of emissions in the use phase presents an opportunity for focused efforts toward reducing those emissions. While the company still seeks energy reductions throughout the supply chain, it has determined that further improvements in the manufacturing process would be hard to find. Bob Karowski, Director Environmental Health and Safety for North America, relates a story from the late 1990's when a group of Enron energy analysts came to evaluate Whirlpool's opportunities for further efficiencies. None were found.

Driven by mandatory (such as national energy efficiency standards) and voluntary (such as Energy Star™) programs, as well as competitive pressure and consumer demand, Whirlpool has been engaged in a constant search for energy efficiencies with its appliances. The company (and the industry) has achieved dramatic energy savings over the past 30 years. Compared to models from 1970, today's refrigerators and dishwashers use approximately one-quarter as much energy and washing machines use approximately one-third as much. Since 1980, the overall percentage of U.S. home energy use that is dedicated to appliances has dropped by two-thirds, to between 18 and 20 percent.

Yet these improvements have not always been easy. In the past, the company has felt that it was paddling upstream against consumer demand. For example, in 1993 the company was the winner of the Super Efficient

Refrigerator Program (SERP) competition sponsored by the U.S. Environmental Protection Agency (EPA), the U.S. Department of Energy (DOE) and 27 national utilities. Though the company received the $30 million prize for winning the challenge and enjoyed the accolades that came with it, some in the company felt that the corporate investment far outweighed the reward. In the end, the prize money barely defrayed the development dollars and the company was forced to go to great lengths to elicit consumer interest in the product. This experience planted concerns within the company that you cannot get too far ahead of the market; efficiency gains must not exceed manufacturing costs or consumer demand.

According to Mark Dahmer, American consumers believed that efficiency was tied to inferior performance. Like the falsehood that higher automobile fuel efficiency necessitates compromised performance, customers believed that an efficient machine would not clean as well. At one point, the situation was so disconcerting that the company engaged in an internal debate over the merits of featuring the Energy Star™ label so prominently on its products. In the end, they decided to keep the label to educate the consumer. While using less water and less energy could elicit concerns from some consumers, the company felt that it had merits as a proxy for quality and performance.

Just after this dampening experience with efficient refrigerators, the company faced a challenge from competitors on efficient washers that, in the end, had a positive effect. In the early 1990's, small European-style front loading, horizontal axis washers were sold in the United States in limited quantities. The sales volume was low, as these products lacked the size or features preferred by consumers. In the late 1990's, the introduction and early consumer acceptance of a new full-sized, front load washer led Whirlpool to rapidly leverage its European technology to introduce a American-style product of its own. This technology was available to Whirlpool through its 1989 acquisition of the Philips business in Europe. The Whirlpool Duet® is a front loading washing machine that uses the more efficient horizontal axis orientation to yield efficiencies of 68 percent less energy, 67 percent less water and 50 to 70 percent less detergent than traditional top loading machines. Most importantly, the machine has been extremely successful in the marketplace and served to counter internal resistance that had been generated by the earlier SERP experience.

Over the past two years, Whirlpool executives sense a market shift as consumers have become increasingly interested in energy efficiency. This, they believe, is driven by both increasing awareness of climate change and environmental issues as well as increasing energy costs. According to Casey Tubman, Brand Manager of Fabric Care Products, "In the 1980s, energy efficiency was number ten, eleven or twelve in consumer priorities. In the last four or five years, it has come up to number three behind cost and performance, and we believe these concerns will continue to grow." But energy efficiency still requires education of the consumer. The most efficient washers can cost up to $500 more than traditional washers (absent any rebates). But, depending on utility rates, they can save between $75 and $100 per year, yielding a payback of five years. According to Catania, "We are getting better and better at making this visible to consumers. This is good for the environment, good for the consumer and good for Whirlpool."

Going forward, Whirlpool believes that the focus on efficiency will have other long-term benefits for the company in terms of market share. According to Tubman, energy efficiency is becoming a source of competitive advantage by building brand loyalty: "Once someone buys a high efficiency device, they never go back to buying a traditional machine." Whirlpool's market research supports this conclusion. According to Steve Willis, Director of Global Environment, Health and Safety, Whirlpool surveys have demonstrated that "there is a strong correlation between a company's performance in appliance markets and their social response to issues such as energy efficiency and pollution." While not uniform across products or regions, Whirlpool believes that environmental attributes (water and energy conservation) yield customer loyalty and repeat purchases.

As an added benefit, Whirlpool executives believe that the company's focus on energy efficiency, like its other responsibility efforts, helps to draw and retain people who feel good about the company and perform better. In Tubman's words, "The values stay here because the people stay here and the people stay here because the values stay here."

All of this leads to the conclusion that a focus on GHG reductions through energy efficiency is central to the company's core strategy. The company states that it will continuously develop new energy efficiency technologies, and at times, license them to competitors. If necessary, the company will also aggressively guard those innovations. In the summer of 2003, Whirlpool sued Korea based LG Electronics for patent infringement, claiming that LG copied technology developed by Whirlpool that delivers sharply higher energy and water savings to customers. When commenting on the suit, David L. Swift, Whirlpool's executive VP for North America, remarked, "Whirlpool has invested heavily in developing innovative fabric care wash technology that delivers meaningful benefits to our customers.... Whirlpool will tirelessly and aggressively work to protect our assets from competitors who choose to disregard U.S. patent law."[156]

The motivation behind the lawsuit, according to Catania, is both to protect the company's assets but also to maintain a level playing field where the company believes that it can win. Toward that end, Whirlpool has worked aggressively through its trade association to develop rigorous techniques and test procedures for measuring energy efficiency, keeping them up to date and uniformly applied.

Organizational Integration

Unlike other companies in this book, the impetus to address GHG emissions at Whirlpool did not originate from the CEO's office. Rather, the ideas were formed in the ranks, then reviewed with the CEO who enthusiastically endorsed the product-based emission reduction target. In the words of CEO Jeff Fettig, "At Whirlpool Corporation we take our environmental responsibilities very seriously. Just as we have taken a global approach to our home appliance business, we believe our world's environmental issues, such as climate change, must be addressed in a similarly comprehensive way. This is why we continue to develop innovative products that minimize their impact on the environment while making our consumers' lives easier."

JB Hoyt, Director of Regulatory and State Government Relations, admits that top down leadership would have been more important at the outset if the company were starting from scratch, but the company had already been working on energy efficiency for years. There was no need to push a new mindset though the organization. In fact, some at the company believe that introducing the concept would do more harm than good, confusing what is already an on-going initiative.

Whirlpool first began attending to climate change in the same way it addresses other environmental issues: through the company's Environmental Council. Comprised of representatives from the six business units (North America, Europe, China, India and Brazil white goods and compressors), the group meets by phone, six to eight times per year to consider the environmental and employee safety concerns facing the corporation. These issues are brought before the Council through suggestions from the Council members based on their efforts to identify best practices, new challenges and emerging trends. In 2003, the Council selected climate change as an issue that necessitated review, particularly in terms of developing better tracking and control of GHG emissions. In addition, the company was motivated by its involvement with the Business Roundtable's challenge to commit to the voluntary Climate RESOLVE initiative. Since Whirlpool's commitment to energy-efficient appliances was central to its long term strategy, the additional focus on GHG emissions was a natural step to take, and had the potential to help create a competitive advantage.

To develop targets for GHG reductions, Willis went to each of the product groups (refrigeration, fabric care (washers/dryers), dishwashers, cooking, air conditioning, and portables) and compiled data on sales volume projections, consumer use, average age of each type of product when taken out of use, and introduction schedules for new, more energy efficient models. He then calculated total energy consumed by all the products over their average life and converted that energy consumption to GHG emissions using country-specific conversion factors. The result was the determination that the total GHG emissions from product sold in 2008 would be three percent less than the total GHG emissions from the product actually sold in 1998.

Willis admits that the three percent goal will not be a tremendous stretch for the company, but according to Hoyt, they may commit to a stretch goal "after we get a track record." Nevertheless, Willis is extremely confident they will meet their target. Returning to the company's mantra of producing efficiency improvements for its consumers, he comments, "We were going to do this stuff anyway. Energy efficiency is one of our priorities." Whirlpool is now working on revisions to both the GHG baseline and the goal in order to reflect the impact of the Maytag acquisition.

One early success story, however, is the impact of consolidating laundry (washer and dryer) manufacturing subsequent to the Maytag acquisition. Whirlpool has consolidated the production from three former Maytag plants into two existing (now expanded) Whirlpool plants. This move resulted in a number of economies of scale, including a 27 percent reduction in emissions from the manufacture of laundry products.

And this leads to one final and important point about Whirlpool's global appliance portfolio. Catania points out that "You'd be hard pressed to tell the difference between a Whirlpool appliance sold in a Kyoto ratified and a Kyoto non-ratified country. We're trying to get as much global leverage on our factories as possible." So, whether the United States ratifies Kyoto or not, the most efficient technologies the company produces (such as seals, the primary source of cooling loss in refrigerators) will migrate around the world. According to Catania, "When we build a factory, we want to milk it and use the technology throughout our product line."

External Outreach

Whirlpool, like other companies in this study, places great emphasis on external outreach. But the direction of that outreach is sharply focused on enhancing consumer awareness and demand. For example, to address consumer misconceptions about the efficacy of energy-efficient appliances, the company has actively worked to educate retailers (such as Lowes and Sears) and consumers on their benefits, including their average five year payback period. Since the more efficient machines need high-efficiency detergents to attain the best cleaning experience (at no additional cost per load, the company is quick to point out), Whirlpool worked closely with Proctor & Gamble to help educate consumers and assure availability of the detergents. Further, the company was pivotal in convincing *Consumer Reports* magazine to include energy efficiency in the rankings of appliances.

Like other companies in this book, the company has not shied away from stepping out in front of its industry on these issues. This has not always been welcomed by its competitors. Early on, the company faced criticism by some who felt that Whirlpool was trying to use energy efficiency as a way to disadvantage the competition, particularly those with lower capital spending plans. This criticism played out with the American Home Appliance Manufacturers (AHAM), the industry lobbying organization in Washington, D.C. of which Whirlpool is the largest member. In 1993, Whirlpool introduced highly efficient refrigerators in the belief that this would spur federal mandates to require manufactures to meet its efficiency level. However, following the Gingrich revolution of 1994, other manufacturers convinced the AHAM to lobby against the new regulations. Since the organization had a policy of one vote per company regardless of market share, Whirlpool's interests were overruled. The organization was successful in convincing the DOE to hold off on the new regulations and, in response, Whirlpool withdrew from the organization in March 1997. After months of negotiations, the company rejoined the AHAM following amendments to the organization's bylaws that require 75 percent membership approval (by market share) of all public policy positions. Today, Catania points out that much of the industry shares Whirlpool's concern for energy efficiency.

Finally, the company has also worked closely with some non-governmental organizations (NGOs) to develop and promote energy efficiency incentives. For example, the company worked closely with the Sierra Club, Natural Resources Defense Council, the Alliance to Save Energy and others to promote manufacturers tax credits within the recently passed Energy Policy Act of 2005. Unlike consumer tax incentives, these credits can offset substantial manufacturer development investments, allowing producers to provide a less expensive, more efficient product to the end consumer. In the words of Tom Catania, the credits provide a "win-win for everyone. NGOs and the government get environmental gains while the consumer gets a better product."

Policy Perspectives

Whirlpool has a long history of working with the government. Since 1975, the company has played a leadership role in crafting every major appliance efficiency regulation, and has been an Energy Star™ Partner of the Year every year except one since 1999. In 2006 and 2007 Whirlpool was recognized with the Energy Star Sustained Excellence Award in recognition of continued contributions in this area.

On the issue of climate change, the company's primary focus on end-use emissions leads executives to feel strongly that any national policy aimed at addressing climate change must include credit for use-cycle reductions. "Who gets the use credits?" asks JB Hoyt. "Should the utility get it? The user? The manufacturer? We're feeling our way along on CSR and climate change. We want to provide a leadership voice." Catania adds, "If the government wants to motivate appliance manufacturers to participate in a meaningful cap-and-trade program, then it needs to provide credit for the power plant emissions reduced or avoided though the increased energy efficiency of our products." Willis echoes this sentiment, "If the company is going to move forward on climate change, we need to get credits for indirect emissions." This is the number one issue, even though the company has been working on emissions reductions for a long time. Says Hoyt, "We would love to get credit for the gains we've made in the 1980s and early 1990's, but the real line in the sand for us is the 1998 baseline for our GHG reduction commitments."

When pressed, Catania adds that the company would be just as satisfied with manufacturer's tax credits rather than carbon credits. In either case, the company will be rewarded for producing energy-efficient products. The least attractive solution, in Catania's view, would be consumer credits for efficient products. "The consumer credit does not have nearly the stimulative effect that the manufacturer's credit." Competitors could easily undercut the stimulus of a consumer rebate by cutting margins.

One final area where Catania has very strong feelings is the topic of state-level climate change regulations. "This would be a huge misdirection of resources and much less would be achieved if we are subjected to a Balkanized set of standards from fifty different sources." In his view, 50 separate policies would benefit neither consumers nor businesses.

Challenges Ahead

Whirlpool is still struggling with the growing pains of recent expansion and acquisitions. Coordination regarding environmental matters among the various divisions of the company is loose. While each plant has an "energy facilities engineer," for example, there is presently no one person in the company who focuses on company-wide energy conservation. (The company is considering creating such a position.) Whirlpool has a highly decentralized culture and its units value and protect their autonomy. Technology sharing is a major thrust of the corporation in areas such as six sigma, process control and lean manufacturing. On environmental matters, there is little technology transfer among plants domestically or globally. These factors have cemented the belief within Whirlpool that the company has reached the limits on energy efficiency in manufacturing process.

An additional challenge is Whirlpool's current difficulty in analyzing emissions data. In order to analyze data collected in 2003 and 2004, the company solicited bids for a data management system to track emissions and conservation. When the proposals came back with costs between $75,000 and $225,000, the company decided to develop a system in-house. These efforts have so far been unsuccessful. According to Willis, a data management system and international GHG conversion factors are the company's biggest current needs with regard to climate strategy.

The company is also alert to the balance it must strike between leading and not leading too much. Consumers care about energy efficiency but cannot be pushed too hard to purchase more efficient models. Requiring sacrifices or greater effort of the consumer so as to attain greater efficiency is out of the question. According to one customer survey conducted by Whirlpool, "Consumers expect a comfortable solution with a minimum of inconvenience. Whoever is the bearer of news to the contrary, is subject to consumer disdain and ridicule."[157]

Looking forward, a focus on energy efficiency gives Whirlpool a premium product well suited for a carbon-constrained future. Though there is relative technological parity between the product offerings of domestic and European manufacturers, the company is concerned that Asian-based manufacturers could overrun the domestic market with cheap, less energy efficient, machines. But increased home energy prices resulting from efforts to reduce GHG emissions could potentially be a windfall to Whirlpool as consumers place an even higher premium on energy efficiency. Banking on this future, Whirlpool has stayed the course and continued to do what it does best—bringing energy efficiency into the home.

Appendix A

A Compendium of Climate-Related Initiatives Used by BELC Companies

This appendix compiles a list of the various climate-related initiatives used by companies in the Business Environment Leadership Council of the Pew Center on Global Climate Change (as of February 2007). The initiatives are organized into seven areas:

- Energy Supply Solutions
- Energy Demand Solutions
- Process Improvements
- Waste Management Solutions
- Transportation Solutions
- Carbon Sequestration and Offsets Solutions
- Emissions Trading, Joint Implementation (JI), and Clean Development Mechanism (CDM) Solutions

---

Energy Supply Solutions

ABB

- ABB manufactures alternative energy and small-scale distributed power generation components and systems that complement existing power markets, including wind farms, fuel cells, and combined heat and power plants using miniature gas turbines. ABB is also developing a number of technologies for energy efficiency and clean energy, including a joint venture with DuPont to develop fuel cell systems.
- ABB has built approximately 1,500 small combined heat and power (CHP, also known as cogeneration) plants in Europe. CHP plants produce both electricity and steam to heat nearby buildings, reducing GHG emissions by 60 percent compared to coal-fired power plants and by about 30 percent compared to natural gas-fired plants.

Air Products and Chemicals

- Air Products' larger hydrogen plants function as "cogeneration" facilities. In addition to producing hydrogen, steam is often produced and exported to a nearby user. The energy efficiency of these hydrogen plants is over 85 percent of what is theoretically achievable, exceeding the 60 percent efficiency level typical of modern natural gas-fired combined cycle turbine power plants.
- A cogeneration unit was also installed to provide energy, heating, and cooling at the Air Products Hersham, UK, European headquarters. This innovative approach for providing energy to an office complex reduced CO_2 emissions by 2700 metric tonnes per year.

Alcoa

- Alcoa and other leading corporations are partnering with World Resources Institute (WRI) to build markets for renewable energy. Convened in 2000, WRI's Green Power Market Development Group seeks to develop corporate markets for 1,000 MW of new, cost competitive green power by 2010.
- Alcoa produces high-efficiency turbine blades for the industrial turbine market in the electric power generation industry.

American Electric Power

- In August of 2004, AEP announced that it is committed to accelerating Integrated Gasification Combined Cycle (IGCC) deployment by building one, or more, commercial-scale, base-load IGCC plants (up to 1,000 megawatts) as soon as 2010. IGCC technology converts coal into a gas and passes it through pollutant-removal equipment before the gas is burned. The process is more efficient and results in fewer emissions of NOx, SO_2 and mercury, in addition to lower carbon dioxide emissions. Carbon capture and sequestration is also expected to be easier and more cost-effective from an IGCC plant than from a pulverized coal plant because the IGCC process creates a high-pressured CO_2 waste stream.
- AEP's second major wind farm, the 160-MW Desert Sky project, was dedicated in May 2002. This project brings the company's total wind generation to over 300 MW, making it one of the largest wind generators in the United States.
- Almost 125 schools participate in AEP's Learning from Light! and Watts on Schools programs, in which AEP partners with learning institutions to install solar photovoltaic systems and to track energy use.
- AEP's Learning from Wind! program provides education on wind power and is used for renewable energy research. Online data from five 10 kW wind turbines allows consumers to monitor real time and historical data on both the local wind conditions and the operation of the turbines, to evaluate whether a small wind turbine might be able to meet their energy needs.
- AEP has constructed a 900 MW state-of-the-art natural gas cogeneration facility for Dow Chemical Company to provide energy and steam to its Plaquemine, Louisiana, site.
- AEP has been co-firing biomass at 4,000 MW of coal-based power generation in the United Kingdom (Fiddler's Ferry and Ferry Bridge) since 2002.
- AEP has begun testing of biomass co-firing at some smaller power plants in its U.S. service territory to evaluate potential reductions in CO_2 and GHG emission levels.

Baxter International

- Baxter switched from using fuel oil to using wood to generate steam in one of its largest manufacturing facilities. The renewable wood fuel is comprised principally of scrap wood from local furniture and lumber operations.
- Baxter installed a photovoltaic system at its Marsa, Malta, manufacturing plant.

CH2M Hill

- CH2M HILL purchases green tags from the Bonneville Environment Foundation for its Portland, Oregon, and Seattle, Washington, offices. In addition, CH2M Hill's corporate headquarters campus in Denver, Colorado, is purchasing 100 blocks (each representing 100 kWh of electricity) of wind power per month from Excel Energy's Windsource program for a period of three years.

Cinergy Corp. (now Duke Energy)

- Cinergy's renewable energy projects launched in 2004 include the donation of a photovoltaic (PV) system for the Cincinnati Zoo's new education center, the installation of a PV array at Cinergy's/PSI Energy's customer service center in Bloomington, Indiana, and the installation of a wind turbine at the Wolcott rest area on Interstate 65 in Indiana.
- Cinergy Corp.'s Cinergy/PSI customers in the state of Indiana have the opportunity to contribute to the Green Power Fund. The fund collects money to be spent on purchasing or developing "green" power from sources such as "low-head" (river areas that have little or no pooling) hydroelectric, wind, and solar photovoltaics. Cinergy/PSI works with the Citizens Action Coalition of Indiana to decide what types of investments the fund will make.

Deutsche Telekom

- Deutsche Telekom installed photovoltaic panels at six sites. Three of them were installed in cooperation with employee training centers.
- Deutsche Telekom's subsidiary DeTeImmobilien installed a 250 kW fuel cell facility in its technical unit in Munich with a goal of reducing emissions by 600 tons of CO_2 per year.
- Deutsche Telekom has forced its energy suppliers to switch to a less carbon-intensive energy mix.

DTE Energy

- DTE Energy is partnering with the U.S. Department of Energy, the state of Michigan, and the city of Southfield to develop, build, and operate a pilot project that will create hydrogen gas from tap water and use that gas in stationary fuel cell generators and to refuel fuel cell vehicles. DTE Energy's Hydrogen Technology Park, a $3-million, five-year pilot project, will be capable of delivering about 100,000 kWh of electricity per year.
- DTE Biomass Energy operates 29 landfill gas recovery projects at sites across the United States. Methane recovered from these projects is converted into pipeline-quality gas, steam, or electricity. DTE Biomass landfill projects have captured the equivalent of more than 25 million metric tons of CO_2.
- DTE Energy's Detroit Edison has promoted geothermal technology in its service area, where nearly 4,000 residential units and two-dozen commercial businesses have geothermal systems.

DuPont

- DuPont and several other companies are partnering with World Resources Institute (WRI) to build markets for renewable energy. Convened in 2000, WRI's Green Power Market Development Group seeks to develop corporate markets for 1,000 MW of new, cost-competitive green power by 2010.
- DuPont announced in June 2005 DuPont™ Generation IV membrane electrode assemblies (MEA) technology for fuel cells that require significantly less catalyst loading compared with the previous generation, while still delivering approximately 20 percent higher power density and well over two times improvement in durability and reliability, leading to more cost-effective fuel cell systems.
- DuPont leads the Integrated Corn-Based BioRefinery (ICBR) project—a U.S. Department of Energy-funded research program. As part of the ICBR, DuPont, the National Renewable Energy Laboratory, and other companies will develop the world's first integrated pilot-scale "biorefinery" that will make use of the entire corn plant—including the stalks, husks, and leaves—to make electricity, biofuels, and an array of biomaterials. For example, in 2003 DuPont received the President's Green Chemistry Award for the development of bio-PDO, a raw material for its Sorona fiber.

- DuPont is partnering with BP to develop "biobutanol," a next generation biofuel that offers significant advantages over biofuels currently on the market. For example, biobutanol has a higher energy content than ethanol and can be transported through the pipelines that make up the existing gasoline infrastructure.
- DuPont is a leading supplier of materials for photovoltaic cells and provides numerous materials for windmills as well.

Exelon

- Selling wind energy in Pennsylvania

  PECO WIND, an Exelon company, is an environmentally friendly power option provided by PECO and leading wind energy marketer Community Energy, Inc., of Wayne, Pennsylvania. PECO launched the product in May 2004, and almost 10,000 customers had enrolled by the end of the year. In aggregate, customers will purchase more than 28 million kWh of wind-generated electricity annually. The environmental benefit is the same as planting about two million trees or not driving 25 million miles.

- ComEd's renewable energy portfolio

  ComEd, a unit of Exelon, purchases electricity generated from landfill methane gas at 22 sites across northern Illinois and wind energy from the 51 MW Mendota Hills project and the 54 MW Crescent Ridge project. In 2004, Chicago passed the 1-MW milestone for installed photovoltaic systems with the completion of the Exelon Pavilions in Millennium Park that integrates photovoltaic into the building's exterior walls—a first-of-its-kind system.

  By the end of 2004, Chicago had more than 50 photovoltaic installations, totaling 1.2 MW. They include systems on ComEd's Chicago South facility, several universities, affordable single-family housing units, and the new Cook County Domestic Violence Court House—at 110 kW, the largest single system in the city to date. The Chicago solar systems represent 86 percent of the solar electric output in ComEd's service territory and 71 percent of the total solar electric output in Illinois, contributing significantly to the state's ranking in the top five.

- Exelon's Wind Generation Portfolio

  Exelon Generation has long-term power purchase agreements (PPAs) with four wind generation projects in Pennsylvania and West Virginia, providing a total wind capacity of 153 MW. The installed capacity associated with these contracts easily makes Exelon the largest wholesale wind marketer east of the Mississippi.

- Financing clean energy in Pennsylvania

  The Sustainable Development Fund (SDF) finances Pennsylvania companies and projects that involve renewable energy, advanced clean energy and energy efficiency technologies. Funded by PECO settlement agreements, SDF is managed by the Reinvestment Fund, a regional nonprofit based in Philadelphia. In addition to providing the environmental benefits of clean energy, SDF helps PECO diversify its power generation options.

  In 2004, SDF approved $4.25 million in production incentives for two wind projects that will add 50 MW of generating capacity in 2005. The incentives are expected to leverage approximately $60 million in private investment. SDF also provided $4.7 million in lease financing in 2003 and 2004 for four energy conservation projects and leveraged $2.8 million from private banks for purchasing participation in these transactions. In 2004, SDF's Pennsylvania Advanced Industrial Technology (PA-AIT) Fund invested $670,000 in three early-stage renewable and clean energy companies. The SDF solar photovoltaic grant program grew to 83 systems and 308 kW of capacity, including PECO's eight solar affordable housing units in Philadelphia. SDF also approved a new round of television and radio spots to encourage support for the PECO WIND product.

Georgia-Pacific

- Georgia-Pacific employs cogeneration in its integrated pulp and paper mills.
- Georgia-Pacific self-generated 158,804 billion British thermal units (Btu) worth of energy at its manufacturing facilities in 2002 using on-site generated or purchased biomass fuels, providing for over 50 percent of the company's energy needs.
- In cooperation with the US EPA and DOE, Georgia-Pacific is applying gasifier technology to full-scale chemical recovery processes at its Big Island, Virginia, mill.

IBM

- IBM and several other companies are partnering with the World Resources Institute (WRI) to build markets for renewable energy. Convened in 2000, WRI's Green Power Market Development Group seeks to develop corporate markets for 1,000 MW of new, cost-competitive green power by 2010.
- IBM's worldwide use of renewable energy accounted for 2.7 percent of its electricity usage in 2005.

Intel

- Intel buys about 14 million kWh of PGE Clean Wind power annually, enough to meet the needs of almost 1,300 average homes in the utility's service territory. Intel Corporation is Oregon's largest retail renewable power user and one of the largest in the West.

Interface Inc.

- Interface facilities use on-site power generation from photovoltaic arrays at the InterfaceFLOR Commercial manufacturing facility in Georgia and the Bentley Prince Street manufacturing facility located in City of Industry, California.
- Interface and several other companies are partnering with the World Resources Institute (WRI) to build markets for renewable energy. Convened in 2000, WRI's Green Power Market Development Group seeks to develop corporate markets for 1,000 MWh of new, cost-competitive green power by 2010.
- Seven Interface facilities now operate with 100 percent renewable electricity, and the overall percentage of renewable electricity used worldwide is 22 percent. As of year-end 2005, 13 percent of Interface's global energy usage was renewable energy primarily sourced through the purchase of Green-e certified RECs, with a small amount of renewable energy from the grid and on-site generation.
- InterfaceFLOR Commercial is a charter partner in the EPA's Green Power Partnership, a voluntary program aimed at boosting the market for power alternatives that reduce the environmental and health risks of conventional electricity generation.
- InterfaceFABRIC uses biomass from waste wood chips to supply 60 percent of its total energy needs in its Guilford, Maine, manufacturing facility.

- InterfaceFLOR Commercial has contracted with the City of LaGrange, Georgia to use landfill gas to replace up to 20 percent of its natural gas usage at the LaGrange, Georgia manufacturing facility.

Ontario Power Generation

- Ontario Power Generation plans to quadruple its green energy supply and invest $50 million in developing renewable energy projects based on wind, solar, small hydroelectric, and biomass by 2005.

PG&E Corp.

- PG&E Corporation's utility in California delivered approximately 7.5 million MWh (10.6 percent of retail electricity sales) from renewable resources in 2002.

Rio Tinto

- Kennecott Energy Company (a Rio Tinto subsidiary) is a member of the FutureGen Alliance that is partnering with the U.S. DOE on FutureGen, a $1 billion project that may lead to the world's first nearly emission-free hydrogen and electricity production plant from coal, while capturing and disposing of CO_2 in geologic formations.
- Rio Tinto's energy product group invests in a number of commercial enterprises and collaborative programs to develop and commercialize new technologies aimed at improving the environmental performance of coal. This includes Pegasus Technologies, a company that uses neural networks to optimize the operation of coal-fired electricity generators, minimizing their fuel requirements and reducing the emission of major pollutants.

Royal Dutch/Shell

- Royal Dutch/Shell's Shell Renewables was established to pursue commercial opportunities in solar, wind, and other renewable energy technologies. The key objective for the solar business is to grow in line with the market, which is currently growing at around 25 percent a year. In the wind business, Shell is focusing on developing and operating wind farms, and selling "green" electricity.
- Royal Dutch/Shell purchased an equity stake in Iogen Energy Corporation in 2002, a world-leading bioethanol technology company. The investment will enable the Canadian-based company to develop more rapidly the world's first commercial-scale biomass to ethanol plant. Iogen utilizes existing agricultural residues such as wheat, oat, and barley straw in its bioethanol process.

SC Johnson

- SC Johnson broke ground for its Landfill Gas Green Energies initiative in April 2003, installing a turbine that produces electricity and steam through cogeneration for its largest manufacturing plant, Waxdale, in southeastern Wisconsin.
- SC Johnson has refined technology to enable cleaner mixtures of landfill gas and natural gas to fuel boilers. This provides over a third of its Waxdale plant's steam energy needs. The company has also installed a 3.2 MW turbine and generator to burn waste methane gas that would otherwise be flared into the atmosphere. SC Johnson burns the methane gas instead of fossil fuels, such as natural gas or coal, to generate electricity and steam for the site's operations, providing about 50 percent of the facility's electricity and 20 percent of its process steam.

Sunoco

- Sunoco's aggregate energy consumption at its five refineries in 2005 was 12.3 percent lower than the 1990 base year. On a normalized basis (million Btus/thousand barrels of crude throughput) the refinery energy usage decreased 23 percent when compared to the 1990 base year.

TransAlta

- TransAlta is expanding its portfolio of renewable energy through its investment in Vision Quest WindElectric, Canada's leading developer of wind power. With TransAlta's investments, Vision Quest has expanded its wind energy portfolio by 400 percent, and expects to continue to grow.
- TransAlta has invested approximately $5 million to build a full-scale demonstration facility for its new clean coal technology. Partnering with two levels of government, equipment providers, and other energy companies, TransAlta hopes to complete the facility by 2010. The technology could reduce the GHG emissions of typical coal plants by up to 80 percent.
- TransAlta is the first in Calgary to service its corporate headquarters through wind-generated electricity. TransAlta also signed a 10-year contract with Vision Quest to supply about eight million kilowatt-hours of electricity annually.

United Technologies

- Pratt & Whitney Specialty Materials & Services (SMS) is a new business redefining entire industries by applying its technologies in unique ways. SMS can now bring significant cost savings to myriad operations with minimal impact on the environment. Waterjet systems, convergent spray technologies and ash cleaning systems are just a few of the innovative ways SMS is saving customers time and money.
- The SHOCKSystem™ solution is being introduced to the domestic coal-fired power generation industry. The SHOCKSystem™ is a combustion-based product that lowers the consumption necessities of coal while still allowing equivalent generation capacity for electricity. Once installed and operational, it lowers fuel consumption requirements while increasing overall boiler efficiency by removing ash and slag deposits that accumulate on boiler tubes. This transaction increases the heat transfer absorption capabilities, thus less heat is lost during the entire power generation cycle.
- ElectroCore™ is a new, advanced power plant emissions control system under development that will control a variety of pollutants from coal-, wood- and other solid fuel-fired boilers, ushering in a new way to control multiple pollutants in power plants and manufacturing facilities.

Weyerhaeuser

- Weyerhaeuser pulp and paper mills supply 70 percent and wood products facilities supply more than 50 percent of their own energy needs through biomass fuels. Weyerhaeuser is also involved in the commercialization of gasification technology that significantly increases the amount of heat and electrical energy obtainable from biomass.
- Weyerhaeuser employs cogeneration in a number of its pulp and paper mills. Its containerboard mill in Albany, Oregon, received EPA's 2005 Energy Star CHP Award for reducing energy and carbon emissions.

Wisconsin Energy Corp.

WEC's emissions in future years will continue to be influenced by several actions planned or underway as part of its *Power the Future* plan, including:

- Repowering the Port Washington Power Plant from coal to natural gas combined cycle units.
- Increasing investment in energy efficiency and conservation.
- Maintaining and increasing the company's nonemitting generation, including renewing Point Beach Nuclear Plant's operating license, potentially adding generation from over 200 MW of wind capacity, and increasing customer participation in the Energy for Tomorrow® renewable energy program.

- Wisconsin Energy's target is to have 5 percent of its Wisconsin retail electric sales come from renewable energy sources by 2011.
- Wisconsin Energy produces or purchases more than 140 MW of renewable energy capacity from a variety of sources inside and outside of Wisconsin. Some of it is used for the corporation's Energy for Tomorrow® residential and commercial renewable energy program, while much of the remainder is used to satisfy the state's Renewable Portfolio Standard.
- In 2005, Wisconsin Energy purchased the development rights for two wind farm projects in Wisconsin. The projects are expected to be on-line in 2008 pending regulatory approval.
- Wisconsin Energy's Energy for Tomorrow® program gives customers the choice of having 25 percent, 50 percent, or 100 percent of their electricity usage charged at a slightly higher rate to support generation or purchase of energy produced by renewable resources. Energy sold through the program in 2005 came from wind (31 percent), landfill gas (65 percent), and low-impact hydro (4 percent). At the end of 2005, 12,458 residential and commercial customers were enrolled in the program, purchasing 53,378 MWh of electricity.

Energy Demand Solutions

ABB

- ABB provides drives, motors, generators, and power electronics designed for the greatest possible efficiency.
- ABB uses life-cycle assessment studies to determine product environmental impacts, including energy use and CO_2 emissions.
- ABB evaluates and reports annually on the environmental performance of its products. An important component of these reports is a life-cycle analysis of the products, including the inputs and outputs of CO_2, SF_6, CFCs, and energy.
- ABB has embarked on a project to develop environmental product declarations for all its major products and services. Environmental product declarations describe the environmental performance of a product, system, or service over its lifetime.

Air Products and Chemicals

- Air Products and Chemicals' efficiency engineers constantly monitor the performance of their major energy-intensive operations. In 2002, those engineers completed numerous global energy efficiency projects resulting in an estimated 26 MW of power savings; this is equivalent to the power consumed by 18,500 average homes annually and equivalent to avoiding 174,000 tons of CO_2 emissions.
- Air Products works closely with its energy suppliers to make the most efficient use of its generation facilities to help minimize greenhouse gas (GHG) emissions. Air Products matches its energy needs to that of the energy supplier by shutting down production at times of peak demand and increasing production at other times. Such efforts contribute to "demand loading," a practice in which energy suppliers try to optimize the generational efficiency of a power plant by ensuring that it runs as close as possible to the point of maximum efficiency.
- Air Products and Chemicals and its partners were selected by the DOE Industries of the Future (IOF) Best Practices Program to demonstrate the potential for using CO_2 to manufacture polyurethane. In addition to using less energy, the new process will be cleaner, significantly reduce the environmental impact of making the foam, and reduce the net release of CO_2.

Alcoa

- Alcoa has reduced the electricity required to produce a ton of aluminum by 7.5 percent over the last 20 years.
- Alcoa supplies lightweight, recyclable materials for motor vehicle assembly; each kilogram of aluminum that replaces higher density materials provides the potential to save 20 kilograms of CO_2e emissions via better fuel economy and recyclability.

American Electric Power

- AEP is implementing energy efficiency plans to offset 10 percent of the annual energy demand growth in its Texas service territory.
- In 2003 alone, AEP invested over $8 million to reduce more than 47 million kWh by installing energy efficiency measures in customers' homes and businesses.

Baxter International

- Baxter's corporate energy management group performs energy reviews of the company's manufacturing facilities, maintains energy use standards, and researches and communicates best practices in energy conservation.
- Baxter holds a worldwide energy conference every two years to share the latest energy conservation technologies and efforts and to recognize individuals and plants for their leadership and accomplishments in saving energy.

CH2M Hill

- CH2M Hill conducted an inventory of energy consumption by over 100 office spaces that the company occupies as a tenant in the United States and Canada. The findings were used to set energy conservation goals in 2004. The company's Denver-area employees moved into a new corporate campus, which was designed using the LEED™ green building system guidelines to reduce energy consumption by 24 percent compared to conventional commercial office buildings.

Cinergy Corp. (now Duke Energy)

- Cinergy's subsidiary, Cinergy Solutions Inc., provides energy analysis and evaluation services and conducts energy efficiency projects. In 2004, Cinergy Solutions announced that it will fund the lighting retrofit of Oldenburg Academy high school to replace fixtures that are 30 to 40 years old.

Cummins Inc.

- Cummins has implemented energy conservation efforts in several of its facilities. Corporate headquarters and other major facilities have agreed to cut electricity consumption by 6 MW on peak demand days. Other facilities have installed air compressor controls and high-efficiency lighting, and have begun using hot water from engine testing to melt snow, reducing the need for electric resistance wiring.

Deutsche Telekom

- Deutsche Telekom achieved the greatest energy efficiency gains in 2002 from optimizing the operation of building technology for switching and transmission technology. By using the waste process heat generated in the operation of IT technology, Deutsche Telekom was able to replace conventional heating systems (e.g., electrical and oil-gas heating systems) in smaller technical buildings. Efficiency gains were also achieved by optimizing air conditioning systems.

- Additional energy efficiency gains were also achieved through the deactivation of digital switching equipment modules that were no longer required. These and other measures contributed to both an increase in energy efficiency and to an impressive reduction in energy consumption of 49.57 GWh in 2002 for T-Com. They have also enabled an emissions reduction of 26,300 tons of CO_2.
- Deutsche Telekom also uses energy optimization teams at its technical branch offices with the goal to develop, analyze, evaluate, and implement energy optimization measures. These teams have been assigned the task of operating a wide range of technology and equipment as efficiently as possible, from complex heating and air conditioning systems, to communications technology, to the energy-related optimization of office workstations.

DTE Energy

- DTE Energy works with customers to find ways they can help protect the environment by using energy wisely. DTE Energy Partnership has a staff of more than 40 energy engineers that work with businesses to increase efficiency.

DuPont

- DuPont used 7 percent less total energy in 2004 than it did in 1990, despite an almost 30 percent increase in production. Compared to a linear increase in energy with production, this achievement has resulted in $2 billion in cumulative energy savings.
- DuPont Tyvek housewrap improves the energy efficiency of buildings, with energy savings in the first year of use alone some 10 to 20 times the energy required to produce the product.
- DuPont engineering polymers used in applications like intake manifolds help to safely reduce the weight of motor vehicles and improve their fuel efficiency.

Entergy

- Entergy, under its commitment to stabilize power plant CO_2 emissions, has implemented 44 internal GHG reduction programs as of December 2003 that will achieve a projected 1 million tons of CO_2 equivalent reduction by 2005. Several of these projects focus on using less fuel to generate electricity at power plants: two projects allow generating units to operate on less power when in stand-by mode, while two other projects are installing advanced controls to regulate the combustion processes in selected plant boilers. These projects are dedicated to improving the efficiency and capacity factor of Entergy's cleanest and lowest emitting fossil, nuclear, and renewable electric generating units.

Exelon

- Exelon Energy Delivery's Smart Returns Load Reduction Programs

  ComEd's 10 Smart Returns products represent one of the largest, most successful loads response portfolios in the United States. The programs provide customers with a financial incentive to curtail use, while benefiting the community and the environment through a lower, more stable load. The more customers curtail use, the more financial incentives they can potentially earn.

  From single-family residential homes to large steel mills, the Smart Returns products provide opportunities for almost every customer to participate. ComEd works closely with large commercial, institutional, and industrial customers to customize curtailment plans and maximize energy efficiency opportunities. On hot summer days, ComEd's load response programs can contribute a 1,000-MW reduction to system peak loads.

  PECO's Smart Returns program has three products, and customers may participate in any or all. The first, active load management (ALM), is a program in which participating customers guarantee that they will reduce their energy consumption within one hour of PECO's request. This emergency program is designed to respond to events that are triggered within the PJM Interconnection, a regional transmission organization. During 2004, PECO gained 26 MW of new ALM customer load, in addition to the existing 74 MW under contract.

Under the second Smart Returns program, voluntary load reduction, customers receive a one-hour notification to curtail energy consumption and share in a percentage of PECO's energy cost savings. PECO had 123 MW under this program for 2004.

Finally, PJM's voluntary Economic and Emergency Load Response Programs provide an additional Smart Returns choice for potential load response customers. During 2004, PECO signed up more than 200 MW.

- EED internal energy efficiency initiative

  The Exelon Environmental Strategy Energy Efficiency Team was charged with the goals of improving energy efficiency at EED facilities by 3 percent annually from 2003 to 2007 and developing recommendations for expanding the program to other Exelon facilities. The team, which supports 74 EED facilities and 8,200 employees, implemented a broad strategy that includes a budget for collateral materials, facility benchmarking and energy audits, efficiency retrofits, and a multiyear communications plan with internal articles, posters and stickers to remind employees to turn off computers and lights when not in use.

  For 2004, EED reduced energy consumption by 7 million kWh compared to the 2002 baseline. Normalizing the data based on 30-year averages and the current year heating and cooling degree-days resulted in an improvement of 4 percent in 2003 and 3.8 percent in 2004, thus exceeding the goal each year.

Georgia-Pacific

- Georgia-Pacific reduced the total energy use in all of its operations by nearly 3 percent between 2001 and 2002, reducing energy consumption from 331,900 billion Btu in 2001 to 326,300 billion Btu in 2002.

Hewlett-Packard

- HP participates in the US EPA's ENERGY STAR® Program, and more than 300 of its products are ENERGY STAR® qualified.
- HP uses "Instant On" technology in many of its laser-jet printers, allowing them to save energy from immediately shifting from active printing to a power saving "sleep mode," without sacrificing printer reliability or the time needed to start the next job.
- HP is developing "all-in-one" products that combine several typical office appliances into one machine, saving up to 40 percent in energy and materials.
- HP has implemented energy-saving measures at many of its own facilities. These measures include installing automated and centralized control systems to minimize energy consumption and maximize efficiency, establishing new temperature set-points, reducing lighting, encouraging employees to turn off lights, computers, and other appliances when not in use, and educating employees about energy conservation. At its Roseville technology campus in California, the percentage of computers left on after work dropped from 33 percent to 8 percent in one year.

IBM

- IBM achieved a 6.1 percent reduction in total conventional energy use through energy efficiency and conservation measures and through the procurement of renewables. This corresponds to an approximate reduction of 173,500 tons of CO_2 at a cost savings of $17.3 million.
- During 1990 to 2002, IBM's energy conservation measures resulted in a savings of 12.8 billion kWh of electricity—avoiding approximately 7.8 million tons of CO_2 and saving the company $729 million dollars in reduced energy costs.

- IBM sets targets for product efficiency for a wide range of products. One-hundred percent of the applicable products first shipped in 2005 met the ENERGY STAR® criteria. In 2005, IBM reported increases in product energy efficiency ranging from 11.5 to 50 percent.

Intel

- Intel, working closely with ENERGY STAR® implemented power management on 65,000 laptop displays and 45,000 desktop monitors worldwide. This initiative will save about 9,650,000 kWh over the next year, or enough electricity to light 11,000 U.S. homes for one month. At $0.05 per kWh, Intel will realize an annual savings of $482,000.
- Intel provides enabling technology for electronics manufacturers to build products that meet or exceed the ENERGY STAR® standard. For example, Intel's Instantly Available PC allows PCs to go to under 5 watts "sleep mode" with wake up in under 5 seconds. From 2002 to 2010, these savings will prevent 159 metric tons of CO_2 emissions.

Interface Inc.

- Interface's improved efficiencies and conservation efforts have reduced the total energy required to manufacture carpet per unit of production by 41 percent since 1996.

Lockheed Martin

- Over 90 percent of greenhouse gas (GHG) emissions at Lockheed Martin derive from energy use. Accordingly, in 2002 the company established an Energy Program within the corporate ESH organization that sets energy management policy, and tracks progress in improving energy efficiency and reducing greenhouse gas emissions. The Energy Program Manager works closely with ESH and facility operations managers across the corporation to promote energy efficiency through employee education, improved maintenance practices, and capital investments. Through the Energy Program, Lockheed Martin has invested over $14 million since 2002 in capital improvements specifically targeted at increasing energy efficiency and lowering greenhouse gas emissions at our operating units.

Ontario Power Generation

- Ontario Power Generation helps customers save energy in their own homes by replacing inefficient appliances, retrofitting lighting, and installing weatherization measures.
- Through its Energy Efficiency Program, Ontario Power Generation has achieved over 2,130 GWh per year in energy savings. Since 1994, OPG has undertaken over 300 thermal conversion efficiency and electrical efficiency projects, saving approximately the amount of energy consumed in a year by a city of 80,000 people.
- OPG has introduced a GreenSaver program to encourage its employees to apply energy efficiency measures at home. OPG gives information sessions on the benefits of home audits for energy conservation.

PG&E Corp.

- PG&E Corporation's utility, Pacific Gas and Electric Company, reduced overall energy use in 2002 at 88 of its California facilities by almost 24 percent compared with 1998 baseline energy usage levels through energy efficiency and conservation. This resulted in savings of almost 28 GWh of electricity, and prevented approximately 7,000 tons of CO_2 from being emitted to the atmosphere.

- Since 1990, Pacific Gas and Electric Company's customer energy efficiency programs have cumulatively saved more than 138 million MWh of electricity (cumulative 36 million to 80 million tons of CO_2 emissions avoided, depending on whether a base or peak load emission factor is used). Customer energy savings realized in 2002 were approximately 4.9 million MWh of electricity and 160 million therms of natural gas—enough to power approximately 740,000 homes for a year. The emissions avoided from these actions alone totaled approximately 2.8 million tons of CO_2.

Rohm and Haas

- Rohm and Haas's largest facility, in Deer Park, Texas, has maintained an aggressive energy management effort and continuously reduced its energy consumption each year since 1997. Energy consumption per pound of product at the site was 26.6 percent lower in 2003 than 1996, eliminating 76,700 tons per year of CO_2 and 1,200 tons per year of NO_x emissions.
- In 1999, the company began engaging its other production facilities in a parallel effort. In 2001 a corporate-wide metrics system was installed to track energy usage on a monthly basis, and a formal goal of year on year reductions (1 percent per pound of product) was implemented. Various efforts at the company's other facilities have included energy assessments, best practice implementation, and process improvements.
- Rohm and Haas's energy-saving products include its DURAPLUS™ roofing system, made with Rhoplex® Emulsion Polymers for reflective roof coatings that can be applied to rubberized roofs to increase the roofing material's life span while lowering the solar radiation to the roof; and SEA-NINE® 211, a biodegradable antifouling agent that prevents biological build-up on large ocean-going vessels, thereby reducing the ship's drag and energy consumption.

Royal Dutch/Shell

- Royal Dutch/Shell is utilizing an in-house developed energy-efficiency program to support its five-year energy-efficiency targets. The program, operated through Shell Global Solutions and known as Energise, helps facilities identify, implement, and sustain energy efficiency projects.

SC Johnson

- SC Johnson reduced volume and temperatures of aerosol water baths, which helped to reduce water bath energy use by 50 percent with no decrease in production quality at its Waxdale aerosol production facility.

Toyota

- Toyota's 624,000 square-foot headquarters expansion in Torrance, CA includes buildings that are expected to exceed state energy-efficiency standards by 20 percent. The facility includes a 500 kW photovoltaic system, and was awarded a certification of LEED™ Gold by the U.S. Green Building Council in April 2003.
- In an effort to reduce energy usage from its sales and distribution network, Toyota established an energy usage database that is updated monthly. Through the help of this database and other efforts, Toyota has reduced total energy consumption by 11 percent in its sales and distribution network since 2000. These savings include the avoided consumption of over 18 million kWh of electricity, 707,000 therms of natural gas, and cost savings of over $2.8 million.

TransAlta

- TransAlta improved its energy efficiency by an estimated 3 to 5 percent by upgrading turbines, cooling towers, advanced control systems, boilers, and heat exchangers.

United Technologies

- UTC Fuel Cells (UTCFC) is a business unit of UTC Power and manufactures the PureCell™ 200 power system, which provides 200 kilowatts of electricity and up to 925,000 Btu/hr of heat for combined heat and power applications. Each PureCell™ 200 avoids the production of 1,100 tons of carbon dioxide emissions annually. In May 2005, the PureCell™ 200 fuel cell fleet achieved a major milestone, providing 1 billion kilowatt hours of energy production, or enough to power 91,000 homes for a year.
- UTC Power has developed the industry's first integrated microturbine and double-effect absorption chiller system, the PureComfort™ 240M. The system converts more than 80 percent of its fuel input to efficient electric, cooling and heating output. It is expected to reduce carbon dioxide emissions by 40 percent and nitrogen oxide emissions by 90 percent over those of the average central fossil fuel generation plant.
- UTC Power, in partnership with Carrier Corporation, developed the PureCycle™ 200 power system to turn waste heat into electricity, providing a zero-emission alternative to traditional power sources. In addition to the environmental benefits, the PureCycle™ 200 offers high reliability, low maintenance, and cost savings through the reduced fuel use.
- Carrier Corporation is the world's leading manufacturer of heating, ventilating, refrigerating, and air conditioning systems and products. In 1994, Carrier pioneered worldwide phase out of CFC's, 16 years ahead of international law for developing countries. Additionally:

  Carrier participates in the EPA's Energy Star program to provide energy efficient products to residents and businesses.

  Carrier was a leading advocate for a national energy policy with a strong commitment to conservation and efficiency improvement, including a consensus energy efficiency standard agreement for commercial packaged air conditioning products, refrigerants and freezers. Carrier was instrumental in moving the industry to a 13 SEER [Seasonal Energy Efficiency Ratio] standard, meaning that Carrier residential air conditioning systems shipped in the United States after January 2006 will be on average 30 percent more efficient than today's standard.

Weyerhaeuser

- In 2004, Weyerhaeuser used 27 percent less energy to produce a ton of product than it did in 1999.

Whirlpool

- Whirlpool manufactures clothes washers, refrigerators, and dishwashers that qualify for the Energy Star label. Some of these appliances exceed U.S. energy efficiency standards by 30 to 50 percent.

Wisconsin Energy Corp.

- Wisconsin Energy's support of demand-side management programs has resulted in reductions of nearly 15 million tons of CO_2-equivalent emissions since 1995.
- In 2004, Wisconsin Energy filed a plan approved by the Public Service Commission of Wisconsin for development of 55.8 MW of energy efficiency by 2008.

Process Improvements

Air Products and Chemicals

- Air Products and Chemicals, working with semiconductor manufacturers, helped to optimize chamber-cleaning processes resulting in perfluorocarbon (PFC) emission reductions of as much as 85 percent.
- Air Products was presented with the 2002 Climate Protection Award from the EPA for its role in reducing PFCs in the semiconductor industry.
- As the world's leading supplier of hydrogen, Air Products is providing hydrogen to petroleum refiners to help them meet government mandates worldwide for producing low-sulfur, cleaner burning gasoline and diesel fuel.
- Air Products has helped pioneer the LNG industry and has been designing liquefaction systems and supplying cryogenic heat exchangers for LNG plants all over the world for the past 30 years.
- For the glass industry, Air Products has enabled yield and efficiency improvements, as well as pollutant emission reduction, from its experience with oxy-fuel technology.
- For the metals industry, Air Products has also provided efficiency improvements via oxy-fuel technology.

Alcoa

- Alcoa's 26 aluminum smelters reduced PFC-generating "anode effects" by 75 percent between 1990 and 2002, resulting in an annual savings of 12 million metric tons of CO_2e.

American Electric Power

- American Electric Power is a charter member of the U.S. EPA's Sulfur Hexafluoride Emissions Reduction Partnership for Electric Power Systems.

Baxter International

- Baxter reduced its process-related GHG emissions by 99 percent between 1996 and 2002 by phasing out the use of high-GWP10 solvents. These process changes resulted in reductions of over 3 million metric tons of CO_2e.

Cinergy Corp. (now Duke Energy)

- Cinergy Corporation is a charter member of the U.S. EPA's Sulfur Hexafluoride (SF6) Emissions Reduction Partnership for Electric Power Systems.

Entergy

- Entergy has replaced electrical equipment containing SF6.

Exelon

- Exelon's ComEd and PECO subsidiaries are members of the U.S. Environmental Protection Agency's Sulfur Hexafluoride Emissions Reduction Partnership for Electric Power Systems.

Intel

- Intel has deployed energy conservation solutions across the company by retrofitting boilers with more efficient Autoflame™ control technology. At Intel's New Mexico site, five boilers were successfully retrofitted at a cost of about $250,000. The return on investment realized was $170,000 per year in natural gas fuel costs, $50,000 per year in electrical energy savings, and $40,000 per year in boiler maintenance costs. Similarly, where the new technology has been installed, there has been an average reduction of nitrous oxide (N_2O) and carbon monoxide (CO) emissions from the boilers of 32 percent and 92 percent respectively.

PG&E Corp.

- PG&E Corporation's Pacific Gas and Electric Company (PG&E) became a charter member of the U.S. EPA's Natural Gas Star Partnership in 1994, and its former subsidiary, National Energy and Gas Transmission (NEGT), joined the program in 2000. Through the systematic replacement of equipment and older pipelines, the company has adopted cost-effective technologies and best management practices to reduce methane losses. Efforts in this area continue to include focused inspections and maintenance at compressor stations, modifying system operations to reduce venting, and reducing frequency of engine restarts with gas. In 2002, PG&E and NEGT undertook numerous activities that resulted in over 185,000 tons of methane avoided. These 2002 emissions avoided equate to over 4.2 million tons of CO_2e.
- PG&E is a charter member of the U.S. EPA's Sulfur Hexafluoride Emissions Reduction Partnership for Electric Power Systems.

Rio Tinto

- Rio Tinto reduced annual GHG emissions by 1.76 million tons compared to business as usual through projects undertaken with the Australian Government's Greenhouse Challenge, a program that helps industry identify opportunities to mitigate emissions.

Royal Dutch/Shell

- Royal Dutch/Shell has ended the practice of continuous venting of gas at oil production facilities and has a target to end continuous operational flaring at such facilities by 2008.

Toyota

- Toyota reduced the energy required to produce a vehicle manufactured in its North American facilities by 7 percent in fiscal year 2002 through process improvements, such as reducing compressed air usage by improving system operating control, and the development of waste heat recovery systems in painting shops.

Wisconsin Energy Corp.

- Wisconsin Energy Corporation is a charter member of the U.S. EPA's Sulfur Hexafluoride Emissions Reduction Partnership for Electric Power Systems. Wisconsin Energy has committed to reducing SF6 emissions to less than 5 percent of its equipment's nameplate capacity. By the end of 2002, Wisconsin Energy reduced annual emissions of SF6 by nearly 95 percent, or to 2.3 percent of total capacity.

Waste Management Solutions

Air Products

- Air Products has successfully reduced the amount of hazardous waste generated per pound of product by more than 50 percent; and reduced air emissions by 60 percent from chemicals facilities that it acquired since 1997.
- Air Products entered into an agreement with a neighboring company to provide the waste stream from one of its dimethylformamide plants for use as a fuel source for that company. This arrangement reduces the neighboring facility's energy demand and lowers the amount of CO_2-forming volatile organic compounds flared by the Air Products facility.
- Air Products has numerous operations that recover hydrogen molecules and other waste gases from the industrial processes of other companies. Hydrogen recovery reduces the amount of natural gas that would otherwise be needed to produce hydrogen.
- Air Products also uses landfill gas to fuel a boiler at one of its operations in Cincinnati, Ohio.
- Air Products Hometown, Pennsylvania plant received the Governor's award for Environmental Excellence for the second time in three years for reducing raw material usage, energy usage and waste generation. Among the achievements were a 1.43 million kWh reduction in electricity usage and 200,000 miles per year reduction in transport miles associated with raw material deliveries and waste transportation.

Alcoa

- Alcoa encourages aluminum recycling by sponsoring recycling programs, operating the Alcoa Recycling Company, supporting research on recycling and alloy separation, and purchasing large amounts of scrap. Aluminum produced from recycled metal requires only 5 percent of the energy required to produce the metal from bauxite ore.
- Alcoa sponsors life-cycle analyses on a number of products, including automotive components, beverage cans, aluminum wheels, and building components, to determine where processes and product designs could be improved.

American Electric Power

- AEP has promoted the use of Coal Combustion Products (fly ash, bottom ash, boiler slag, and flue gas desulfurization scrubber materials) since the early 1950s. In 2002 alone, AEP sold over 1.3 million tons of CCPs, utilized over 1.1 million tons for internal projects, and donated another 161,000 tons. In all, over 32 percent of CCPs were utilized, avoiding the use of substantial amounts of landfill space.

Baxter

- Baxter reduced its generation of nonhazardous waste by 14 percent per unit of production value between 1996 and 2002 and recycled 47.3 million kilograms of waste in 2002 alone. Baxter also reduced the amount of packaging used per unit of production by 15 percent between 1995 and 2002, with a 3.7 percent reduction in 2002 alone (as compared to a 1995 baseline), a reduction of 1.8 million kilograms. Baxter's 2002 reduction alone has saved the company $2.9 million dollars.

California Portland Cement

- California Portland Cement Company re-uses industrial waste as a raw material for clinker (a product used in cement production), eliminating the need for land disposal and reducing GHG emissions.

CH2M HILL

- CH2M HILL surveyed its 100-plus U.S. and Canadian offices about sustainable office management practices. The 2003 survey results showed that 95 percent of offices are recycling white paper, and more than 86 percent are recycling aluminum cans and toner cartridges.

Cinergy Corp. (now Duke Energy)

- Cinergy Corporation has implemented an extensive program for the reuse of fly ash, a by-product of coal combustion. This significantly reduces the volume of materials that require land-filling, and provides a substitute for more energy-intensive materials.
- Cinergy has implemented recycling programs in its offices (for paper) and generating plants (for metals).

Cummins Inc.

- Cummins' ReCon program facilitates the reuse and recycling of Cummins diesel and gasoline engines and components. Through the program, Cummins remanufactured 25,000 engines and over 1,000,000 diesel components in the year 2000. Each year, ReCon plants also generate approximately 3,000 tons of scrap metal for recycling.
- Through a voluntary recycling program, employees at Cummins' San Luis Potosi facility were able to save the equivalent of over 9000 seven-year-old trees and over 2 million kWh of electricity.

Deutsche Telekom

- Deutsche Telekom set a target to reduce its waste materials for disposal by 3 percent annually during 2001–2004. Deutsche Telekom reduced its waste for disposal by approximately 39 percent between 1995 and 2000.

DuPont

- The DuPont-Solae plant in Memphis, Tennessee, uses landfill gas as a replacement for natural gas to fuel boilers and other plant equipment, replacing more than 90 percent of the natural gas used by the site's boilers. The U.S. EPA has calculated that area greenhouse gas emissions have been reduced by an equivalent of the removal of 70,000 cars from the road or planting 95,000 acres of forest.

Entergy

- Entergy recycles over 70 percent of its power plant waste ash. The majority of the ash is utilized in the production of concrete. This reduces the volume of material sent to landfills and reduces the energy requirements and CO_2 emissions associated with the processing of materials traditionally used to produce concrete.
- Entergy has funded a project in the eastern United States that will collect coal mine methane vented from abandoned mines and convert it to pipeline-quality gas or use it as fuel to generate electricity. The project will reduce GHG emissions by 400,000 metric tons of CO_2e through 2005.

Exelon

- Landfill Gas to Energy

  Exelon continues to reduce overall greenhouse gas emissions by supporting landfill gas to energy recovery. Utilizing landfill methane to generate electricity produces less environmental impact than burning fossil fuels and has the added benefit of capturing an energy source that otherwise would have gone to waste.

Exelon Power is in the final year of a two-year project to convert an oil-fired plant designed in 1950 into a 21st century, clean operating, reliable and efficient generating station through the use of improved technology and production methods. As a result of this project, the two-unit 60 MW Fairless Hills Generating Station will be the second-largest landfill gas generating station in the U.S.; a substantial renewable energy project able to consume 100 percent of the landfill gas that Waste Management produces at its nearby GROWS and Tulleytown landfills; and a significant contributor to Exelon's greenhouse gas reduction target through its consumption of landfill gas that would otherwise have been flared.

Exelon Power also operates the 6 MW Pennsbury plant in southeastern Pennsylvania that utilizes landfill gas to generate electric power. Exelon Power was awarded a 1997 Governor's Environmental Excellence Award for its landfill gas projects.

In addition, ComEd purchases electricity generated from landfill methane gas at 22 sites across northern Illinois. To date, Exelon landfill gas initiatives have avoided over 21 million CO_2-equivalent tons of emissions.

- Coal combustion product reuse

  Exelon continues its commitment to reuse the by-products of coal combustion at its fossil generating stations—fly ash, bottom ash, basin ash, and flue gas desulfurization products—and prevent them from consuming valuable local landfill capacity. The company uses these materials for applications that include restoration of land contours at coal mine reclamation sites, antiskid agents for icy roads, production of fertilizer products, and waste-stabilization media.

  In 2004, Exelon continued its commitment to reuse the large volume of products that result from burning coal. The first year that 100 percent of the fly ash, bottom ash, and basin ash and scrubber products were kept out of local landfills was 2002. That accomplishment included more than 137,000 tons of ash materials and just over 21,600 tons of by-products from the SO_2 scrubbing process. Greater demand for power in 2003 increased that challenge to 175,700 tons of ash products and 28,800 tons of scrubber by-products, and the challenge was met.

Georgia-Pacific

- Georgia-Pacific used 68 percent of waste it generated in 2002 for a beneficial purpose.
- Georgia Pacific's North American facilities' formal waste reduction programs have avoided more than 8 million tons of waste since the program began.
- In 2002, Georgia-Pacific used 3.2 million tons of recovered paper to make its products and produced 1.5 million tons of 100 percent recycled content fiber for packaging and tissue products. In 2002, 16 percent of the total wallboard produced was reclaimed from a variety of sources including waste wallboard from construction sites, off-spec wallboard from production processes, and by-products of other industrial processes.
- In 2002, 5.1 tons of wood waste per thousand cubic feet of product manufactured were reused by Georgia Pacific's building products manufacturing facilities (including plywood, oriented strand board, and industrial wood product panels and lumber). This number is up from 3.9 tons in 2001.

Hewlett-Packard

- HP designs its products with recyclability in mind. It operates end-of-life recycling programs for its hardware products in 16 countries and offers toner cartridge recycling programs internationally to 90 percent of the cartridge market.

Holcim (US)

- Holcim is working within existing material specification standards to replace cement clinker with mineral components such as fly ash, a waste material from coal-burning electric utilities, and slag, a waste by-product of steel manufacturing. Each ton of clinker eliminated avoids 1 ton of CO_2 emissions that would have resulted from its manufacture. The company has eliminated 400,000 tons of CO_2 as a result of clinker-factor reductions and is working with government agencies and construction material specifiers to encourage further use of lower clinker-factor cements in concrete.

IBM

- In 2005, IBM's hazardous waste generation indexed to output was reduced 19 percent. This means that the source reduction efforts reduced the generation of haz-waste by 847 metric tons.
- Over the past five years, IBM's total hazardous waste has decreased by 35 percent, and has decreased by 94.5 percent since 1987.
- IBM recycled approximately 78 percent of its non-hazardous waste and 54 percent of its hazardous waste in 2002.

Intel

- Intel recycled 59 percent of its hazardous waste generated worldwide and 73 percent of its solid waste generated worldwide in 2003.
- Additionally, paper with 30 percent recycled content was purchased for all its U.S. copiers and printers.

Interface Inc.

- Interface has reduced its carpet and textile solid waste sent to landfill by over 34 percent since 1996 through waste reduction programs and expanded recycling and reuse programs.
- Interface also uses postconsumer materials otherwise destined for landfills, to manufacture its products, (e.g., soda bottles are used to manufacture Terratex® fabric).
- Interface also accepts postindustrial and postconsumer carpet and diverts it from the landfill using it for product or waste to energy. In 2004, Interface diverted 17 million pounds of carpet from landfills.

SC Johnson

- SC Johnson has achieved more than a 90 percent recycling rate across its global operations since 1990. In 2003, SC Johnson reduced combined air emissions, water effluents, and solid wastes per kilogram of product produced by over 15 percent compared to its year-2000 baseline.

Toyota

- Toyota has reduced the amount of hazardous waste going to landfills from its plants by 40 percent since 2000 and its nonhazardous waste by 11 percent.
- In 2003, Toyota implemented a nationwide, Web-based waste tracking system to better collect and analyze waste-related data to enable further reductions throughout Toyota's North American manufacturing and distribution operations.
- Toyota is also increasing the use of reusable packaging in shipments to distributors.

TransAlta

- TransAlta, along with Ontario Power Generation, has contributed to carbon emissions reductions of over 20,000 metric tons by selling flyash to regional concrete and cement producers.

United Technologies

- United Technologies reduced its domestic hazardous waste production by 41 percent between 1999 and 2002.
- The company set a goal in 1998 to reduce recycled waste by 35 percent and non-recycled waste by 60 percent by 2007.

Weyerhaeuser

- Weyerhaeuser collected for recycling more than 6.7 million tons of paper in 2004, approximately 13 percent of the paper recovered in the U.S. and enough to fill more than 130,000 freight cars. Typical recyclables include old corrugated containers, office wastepaper, old newspapers, and printing papers. More than 4 million tons of the recycled material Weyerhaeuser collects is used in its mills to make new paper. The rest is sold to customers around the world. Recycled fiber comprises about 35 percent of the content of new Weyerhaeuser paper, as averaged across all grades of paper produced by the company.

Wisconsin Energy Corp.

- Wisconsin Energy Corporation utilizes fly ash, municipal wastewater, and paper mill sludge to produce a patented construction product, replacing fossil fuel generation and reducing the amount of solids placed in landfills. In 2002, the company beneficially used 96 percent of these combustion products, compared to a national average of 31.5 percent in 2001.

Transportation Solutions

Air Products and Chemicals

- Air Products and Chemicals' distribution fleet is over 50 percent more fuel-efficient than it was three decades ago. Air Products uses sophisticated logistics scheduling software to maximize the amount of product hauled in each load and determine the optimal delivery routes to customers. Air Products fleet managers have recently set new internal miles per gallon targets to increase fleet efficiency using best practices for driving and maintaining vehicles.
- Air Products develops hydrogen infrastructure and fuel-handling technologies to enable the commercialization of hydrogen as an energy carrier and is working with the private and public sectors to develop a market for hydrogen fuel.
- Air Products is providing hydrogen production, distribution, and vehicle expertise to collaborations of public, private, and government institutions, and is participating in numerous demonstration projects in North America and Europe on the development of hydrogen fuels, fueling systems, and vehicles.
- Air Products is part of the California Fuel Cell Partnership, a unique collaboration of auto manufacturers, energy companies, fuel cell companies, and government agencies. The partnership's goal is to advance and evaluate new automobile technology that can move the world toward practical and affordable environmental solutions. The organization was formed in April 1999 and placed over 40 fuel cell vehicles—cars and buses—on the road between 2000 and 2003. In addition to facilitating the placement of up to 300 vehicles in fleet demonstrations between 2004 and 2007, partnership members will build demonstration hydrogen fuel stations, act to facilitate a path towards commercialization of hydrogen, and enhance public awareness and support.

Baxter International

- Baxter encourages facilities to establish and offer ride-sharing programs to conserve the amount of fuel employees use to travel to work.
- Baxter estimates and reports energy-related greenhouse gas emissions associated with employee commuting and employee use of commercial airlines.
- Baxter has partnered with the University of Puerto Rico in Mayaguez in a pilot biodiesel project. A Baxter vehicle uses biodiesel produced by a system developed at the University. Baxter plans to expand its fleet of biodiesel vehicles in Puerto Rico.

Boeing

- Boeing is involved in a demonstration project aimed at exploring the use of fuel cell technology for future aerospace applications. The project will evaluate the potential application of fuel cell technology to future commercial airplane products. As part of the evaluation, the project will develop and demonstrate the use of fuel cells in auxiliary power units.

BP

- BP is part of the California Fuel Cell Partnership, a unique collaboration of auto manufacturers, energy companies, fuel cell companies, and government agencies. The partnership's goal is to advance and evaluate new automobile technology that can move the world toward practical and affordable environmental solutions. The organization was formed in April 1999 and placed over 40 fuel cell vehicles—cars and buses—on the road between 2000 and 2003. In addition to facilitating the placement of up to 300 vehicles in fleet demonstrations between 2004 and 2007, partnership members will build demonstration hydrogen fuel stations, act to facilitate a path towards commercialization of hydrogen, and enhance public awareness and support.
- BP's Global Choice program allows Australian business customers to offset the greenhouse gas emissions from their fuel consumption. Participation in the program is free for companies purchasing BP Ultimate or bp autogas and only 1–2 cents per liter to offset regular unleaded or diesel fuels. The offsets are independently audited and certified by the Australian Federal Government's Australian Greenhouse Office (AGO). Since November 2001, over 6,500 customers have offset 626,095 tonnes of greenhouse gases.

CH2M Hill

- CH2M Hill supports local commuter trip reduction programs by partially subsidizing the cost of annual transit passes and enabling full-time teleworking for employees whose jobs permit.
- The Denver, Colorado, office initiated a Web-based commuting tool to help employees avoid congestion resulting from a major transportation construction project, to connect people to carpools, and to provide local transit information and traffic updates.
- Some local offices include hybrid gas-electric vehicles in their fleets or subscribe to FlexCar programs.

Cinergy Corp. (now Duke Energy)

- In 2004, Cinergy announced plans to purchase five Toyota Prius hybrid vehicles for its corporate fleet. CO_2 reductions from the use of the five Priuses will be an estimated 37,140 pounds annually (as compared to operating five standard vehicles in Cinergy's corporate fleet).
- Cinergy Corp. operates passenger vehicles, light trucks, and heavy trucks that utilize compressed natural gas or propane as an alternative to gasoline or diesel fuels.

Cummins Inc.

- Cummins joined the U.S. government and other industry partners in the Twenty-First Century Truck Initiative, with the goal of developing commercially viable truck and propulsion system technologies that will dramatically cut fuel use and emissions from medium and heavy-duty trucks and buses.
- Cummins sold over 2,000 Compressed Natural Gas engines to the Beijing Public Transportation Corporation for the city bus fleet. These engines exceed Euro II emissions standards.

- Cummins has partnered with Lockheed Martin Control Systems and Orion Bus to produce the diesel engine and soot filter for Lockheed's hybrid electric drive system for 125 Orion VII hybrid buses, to be purchased by the New York City Metropolitan Transit Authority.

Deutsche Telekom

- Deutsche Telekom's vehicle fleet's CO_2 emissions have decreased approximately 30 percent over the last six years, as a result of using smaller or alternative-fuel vehicles, substituting train travel for car or plane travel, using videoconferencing in place of travel, and incorporating environmental impacts into the company's technical specifications for vehicle suppliers and manufacturers.

DuPont

- Pioneer Hi-Bred International, Inc., a DuPont company, the world's leading developer and supplier of agricultural seeds, operates a significant portion of its fleet of farm and transportation equipment on biofuels such as ethanol and biodiesel, offsetting CO_2 emissions from fossil fuels.

Exelon

- ComEd continues to be a major voluntary user of B-20 biodiesel blended product, with 2004 consumption surpassing the 2 million gallon mark. For 2004, this consumption level reduced particulate emissions by more than 340 tons and displaced the need to purchase more than 400,000 gallons of petroleum-based diesel. ComEd is recognized as the largest regional consumer of biodiesel and ranks in the top 5 percent of biodiesel consumers nationwide.
- In 2005, Exelon purchased 50 Ford Escape Hybrids, the first production hybrid sport-utility vehicle (SUV). These now comprise about 25 percent of the company's overall SUV fleet. The hybrid Escapes provide an estimated 50 percent improvement in city/highway fuel economy when compared to the conventional Escape.
- In 2004, Exelon also joined the Hybrid Truck Users Forum (HTUF), a project of the U.S. Army and WestStart. The forum coordinated specifications and a request for proposal (RFP) for the prototype of a medium-duty hybrid utility truck.

Hewlett-Packard

- In 2006, Hewlett-Packard was ranked in the top twenty Fortune 500 companies participating in the public-private sector voluntary program Best Workplaces for Commuters.℠ Best Workplaces for Commuters℠ was established by the DOT and EPA to publicly recognize employers whose commuter benefits address parking, congestion, and environmental impacts associated with driving-alone commuting. 70 percent of HP's employees telecommute on a full-time, regular, or occasional basis. HP's Bay Area work sites also have electric vehicle recharging stations on-site and offer transit subsidies to employees. HP work sites in Georgia also offer transit subsidies, hold quarterly meetings to discuss commuter issues, and subsidize all vanpool expenses beyond the cost of gas.

IBM

- In 2006, IBM was ranked in the top twenty Fortune 500 companies participating in the public-private sector voluntary program Best Workplaces for Commuters®. Commuter programs, particularly telecommuting, not only benefit the environment by reducing traffic congestion but also benefit IBM employees by providing them with greater flexibility and benefit the company by enhancing the productivity of its work force. IBM has currently more than 20,000 employees participating in telework arrangements in the U.S. Many IBM locations around the country also encourage employees to take public transportation, carpool, vanpool, use bikes, etc., in order to reduce traffic congestion and its resulting air pollution. At these multiple locations, IBM provides commuter assistance programs that provide employees with guidance on using alternative modes of transportation and Emergency Ride Home programs. Some of these IBM locations provide employees with various benefits including but not limited to transit subsidies, discounted transit passes, internal carpool ride-matching service, access to on-site amenities such as cafeterias, credit unions, ATMs, medical center, commuter information kiosks, common telework stations, bike racks, showers, etc.

Intel

- In 2004, Intel was ranked number one among Fortune 500 participating companies in the public-private sector voluntary program Best Workplaces for Commuters℠. In 2003, 44 percent of Intel's 48,600-plus U.S.-based employees were able to take advantage of telecommute options, while other staffers participated in flextime, compressed workweeks, part-time hours, and job-share programs. Intel offers commuter benefits to more than 90 percent of its work force including a universal vanpool and transit subsidy program and Emergency Ride Home services. In addition, Intel provides on-site fitness centers, food cafes, dry cleaning, and photo developing for its employees at major work sites.

Interface Inc.

- Interface addresses transportation impacts by reducing packaging materials, manufacturing its products closer to the customer, and transporting information rather than matter using electronic and on-line communication.
- Interface is a charter partner in the EPA's SmartWay Transport voluntary partnership. SmartWay focuses on reducing pollution and greenhouse gas emissions from ground freight carriers.
- Interface has partnered with Business for Social Responsibility's Green Freight Group.
- Interface has launched its own Transportation Working Group comprised of representatives from all of its business units. The company is working to establish its transportation footprint, including setting a baseline year, developing metrics to monitor future performance, and collecting and sharing best practices between business units.

Ontario Power Generation

- Ontario Power Generation has committed to lease 10 low emission vehicles, electric, gas/electric hybrid or natural gas, each year for the next three years.

PG&E Corp.

- PG&E Corporation began its Clean Air Transportation program in 1988 and currently has more than 650 natural gas vehicles in its fleet.

Rio Tinto

- Rio Tinto subsidiary US Borax is participating with Millennium Cell in the further development and possible commercialization of a process that generates pure hydrogen or electricity from environmentally friendly raw materials such as borates. In the Hydrogen on Demand™ process, the energy potential of hydrogen is carried in the chemical bonds of sodium borohydride, which in the presence of a catalyst either releases hydrogen or produces electricity.

Royal Dutch/Shell

- Royal Dutch/Shell is part of the California Fuel Cell Partnership, a unique collaboration of auto manufacturers, energy companies, fuel cell companies, and government agencies.
- Royal Dutch/Shell's Shell Hydrogen was established in early 1999 to pursue and develop global business opportunities related to hydrogen and fuel cells. Shell Hydrogen is involved, through Icelandic New England Ltd, in a pioneering project that may bring about a complete transition to a hydrogen economy in the coming decades in Iceland.

SC Johnson

- SC Johnson offers a Van Pool program to employees who commute everyday from Milwaukee or Chicago to its global headquarters in Racine, Wisconsin.

Toyota

- Toyota is developing and producing clean energy vehicles, including hybrid, electric, compressed natural gas, and fuel cell electric vehicles. The Toyota Prius, a gas-electric hybrid, became available in the United States in June 2000. Through September 2004, Toyota sold more than 100,000 hybrid Priuses in North America. In 2005, Toyota released two hybrid SUV models, the Lexus RX 400h and the Highlander Hybrid. Toyota has established a goal of selling 300,000 hybrid vehicles annually around the world by mid-decade.
- Over the past decade, Toyota's new automobile fleets have consistently achieved higher fleet average fuel economy than both the industry standard and the Corporate Average Fuel Economy (CAFE) standard required by U.S. law for both car and nonpassenger (light truck and SUV) fleets.
- Toyota is part of the California Fuel Cell Partnership, a unique collaboration of auto manufacturers, energy companies, fuel cell companies, and government agencies.
- Toyota is also part of the Canadian Transportation Fuel Cell Alliance (CTFCA) a public/private initiative to demonstrate and analyze fuel cell fueling options for fuel cell vehicles in Canada.

United Technologies

- UTCFC is developing zero emission, energy efficient fuel cells for transportation applications with environmental and energy security benefits. UTC has deployed zero-emission fuel cell buses in Washington, DC, California, Madrid, and Turin. In 2004, AC Transit logged over 8,000 miles operating a Thor 30 hydrogen fuel cell, hybrid-electric bus developed by ISE Corporation and UTC Fuel Cells. This bus was deployed in the Oakland, California, area and achieved double the fuel economy of a 30-foot diesel bus. In 2005, UTCFC is delivering power plants for four fuel cell buses that will be operated in California by AC Transit and SunLine Transit.
- UTCFC is currently working with major automobile manufacturers, including Nissan, Hyundai-KIA, and BMW and the U.S. DOE on development and demonstration programs for automobiles. They teamed with Chevron and Hyundai-KIA as part of DOE's Hydrogen Learning Demonstration Program and will be deploying a fleet of 32 zero-emission Hyundai-KIA Tucson sport-utility vehicles and Sportage cars as part of the initiative.

Wisconsin Energy Corp.

- Wisconsin Energy Corporation has natural gas vehicles representing approximately 4 percent of the company's light-duty fleet. Since 1999, We Energies has operated the largest NGV fleet in Wisconsin. In addition, Wisconsin Energy has worked with other businesses to place over 700 natural gas vehicles in service and has been instrumental in siting 13 public-access and 9 private compressed natural gas stations.
- In 2005, Wisconsin Energy Corporation was named a Best Workplace for Commuters by the U.S. Environmental Protection Agency for offering its employees outstanding commuter benefits and meeting the EPA's National Standard of Excellence. Offered options include pretax benefits for parking and transit users, ride share program for matching carpool participants, commuter value programs (value passes and commuter coupons), and a bicycle program including bike racks, lockers, and pumps. More than 300 of the 1,400 employees who work in the downtown Milwaukee headquarters participate in these programs.

Carbon Sequestration and Offsets Solutions

ABB

- ABB built the world's first commercial CO_2 capture facility at its Shady Point, Oklahoma, coal-fired power plant. It captures 200 tons of CO_2 a day from the plant's flue gas, which is purified, liquefied, and sold to the food products industry.
- In conjunction with Pacific International Center for High Technology Research and the Natural Energy Laboratory of Hawaii Authority, and other research facilities in Japan, the United States and Norway, ABB is studying the possibility of storing CO_2 in the ocean floor.

Air Products and Chemicals

- Air Products is a technology developer and provider for the CO_2 Capture Project (CCP), which is an international effort by seven of the world's leading energy companies. This project seeks to develop new technologies to reduce the cost of capturing CO_2 from combustion sources and safely storing it underground. It is a collaborative effort involving partnerships with governments, industry, NGO's and other stakeholders.
- Under the CCP, Air Products has directly contributed to projects, including early development of sorption enhanced water gas shift process and advancing the feasibility study of retrofitting boilers and fired heaters with oxy-fuel burner systems.
- Air Products is also participating as a partner in the CANMET programs exploring means to increase efficiency in energy-intensive industries, develop more efficient hydrocarbon conversion processes, and reduce emissions, including CO_2, from fossil fuel combustion.

Alcoa

- Alcoa plants thousands of trees annually near their operations and service areas, sequestering thousands of tons of CO_2 every year.
- In 2003, Alcoa employees fulfilled a goal to plant 1 million trees around the world in ten years—and did so in half the time. A new company goal is for employees to plant 10 million trees by the year 2020.

American Electric Power

- AEP is participating in forest conservation and reforestation projects. By replanting degraded areas, or protecting land that would otherwise be logged, the company is helping to sequester millions of tons of carbon.
- AEP, under DOE's Climate Challenge Tree Planting Project, has planted 21,914 acres with nearly 19 million mixed hardwood and conifer trees at a cost of approximately $5.7 million. Projected CO_2 sequestration is 4.7 million metric tons over the term of the project. In a separate initiative in Louisiana, AEP has planted 9,784 acres with nearly 3 million bottomland hardwood trees at a cost of $6.25 million. Projected carbon sequestration is over 4.4 million metric tons.
- AEP is a founding member of PowerTree Carbon Company, LLC, a voluntary carbon sequestration initiative. PowerTree, which has 25 member companies, will invest $3.4 million for reforestation of over 3,800 acres of bottomland hardwood projects in Arkansas, Mississippi, and Louisiana. The project will sequester over 2 million tons of CO_2 over the 100-year project term.
- AEP's Mountaineer Plant is the site for a $4.2 million carbon sequestration research project funded by the DOE and a consortium of public and private sector participants. Scientists from Battelle Memorial Institute lead this climate change mitigation research project, which will also involve researchers from several other partnering organizations and universities. This project is obtaining the data required to better understand the capability of deep saline aquifers for storage of CO_2 emissions from power plants.

- AEP (a Rio Tinto subsidiary) is a member of the FutureGen Alliance that is partnering with the DOE on FutureGen, a $1-billion project that may lead to the world's first nearly emission-free hydrogen and electricity production plant from coal, while capturing and disposing of CO_2 in geologic formations.
- AEP is a participating member of the UtiliTree Carbon Company. UtiliTree is a consortium of 41 utilities organized by the Edison Electric Institute to invest in a portfolio of forestry projects that manage GHG emissions, particularly CO_2. A $3.2-million investment in eight domestic and two international projects will capture over 3 million tons of CO_2 over the life of these projects.
- AEP is working with the MIT Energy Laboratory as part of a consortium researching the environmental impacts, technological approaches, and economic issues associated with carbon sequestration. The MIT research focuses on efforts to better understand and reduce the cost of carbon separation and sequestration.

BP

- BP is a member of the CO_2 Capture Research Project (CCP), an international effort by seven of the world's leading energy companies. BP has been capturing and storing about 1 million tons of CO_2 per year from a natural gas processing plant in Algeria since 2004.
- BP is working with the MIT Energy Laboratory as part of a consortium researching the environmental impacts, technological approaches, and economic issues associated with carbon sequestration.
- BP is contributing to the development of a Blue Chip Standard, as part of the Climate and Biodiversity Alliance, for demonstrating the contribution of forestry projects to the goal of atmospheric greenhouse gas stabilization. This standard will support the creation of carbon sequestration credits that are generally recognized and therefore tradable.

Cinergy Corp. (now Duke Energy)

- Cinergy is funding the purchase of trees for a 300-acre reforestation project being managed by the Nature Conservancy in Indiana. The project will sequester approximately 75,000 tons of carbon dioxide annually.
- Cinergy Corp. has developed partnerships with various conservancy groups, such as the Nature Conservancy, Ducks Unlimited, and the National Wild Turkey Federation, to plant trees to restore lowland wetlands and riparian zones along rivers and streams and to reforest marginal agricultural lands.
- Cinergy is a partner in the Rio Bravo Carbon Sequestration Project to protect 65,000 acres of endangered rainforest in Belize. The project combines land acquisition and sustainable forestry and is expected to sequester approximately 2.4 million metric tons of carbon over 40 years. Cinergy, the Nature Conservancy, and the Belize Government entered into an agreement to transfer the carbon offsets from the project to Cinergy Corp. The agreement was reviewed and approved by the U.S. Initiative on Joint Implementation (USIJI).
- Cinergy is a participating member of the UtiliTree Carbon Company.

DTE Energy

- DTE Energy seeks opportunities to sequester carbon dioxide and capture methane escaping from landfills. Since 1995, DTE Energy has planted 20 million trees in Michigan alone and DTE Biomass landfill projects have captured the equivalent of nearly 20 million tons of CO_2.
- DTE Energy is a participating member of the UtiliTree Carbon Company.
- DTE is a founding member of PowerTree Carbon Company, LLC, a voluntary carbon sequestration initiative.

Entergy

- Entergy plants thousands of trees annually on their landholdings, sequestering thousands of tons of CO_2 every year.
- Entergy in partnership with Trust for Public Land and the U.S. Fish and Wildlife Service (USFWS), is acquiring 1,600 acres of land adjacent to the Tensas River Wildlife Refuge, restoring bottom land hardwood habitat on marginal croplands and donating the improved land to USFWS who will manage the property. This will sequester 640,000 tons of CO_2 over the next 70 years.
- Entergy in partnership with the Conservation Fund, USFWS and Friends of the Red River, dedicated the Red River Wildlife Refuge in Natchitoches, Louisiana, and established a 600 acre sequestration site that will create 225,000 tons of CO_2 offset credits over the next 70 years.
- Entergy has leased 30,000 tons of CO_2 offset credits from the Pacific Northwest Direct Seed Association (PNDSA). Credits are generated by growers who have agreed to use direct seed agriculture methods for at least 10 years. Direct seed cultivation avoids soil losses from oxidation associated with traditional farming techniques and also reduces the growers' fuel use and soil erosion.
- In December 2003, Entergy became the first U.S. utility to purchase carbon emissions credits from geological sequestration projects. These projects capture CO_2 vent gases that would otherwise be released into the atmosphere and then place them into oil-bearing geologic formations for use in enhanced domestic oil recovery. Entergy is a member of the Gulf Coast Carbon Center and is looking to demonstrate carbon capture technologies, conduct research into geologic sequestration monitoring and verification, and looking to develop an infrastructure in the Gulf Coast region to utilize anthropogenic CO_2 for enhanced oil recovery. As of May 31, 2006, Entergy had purchased 1,500,000 emission reduction credits from enhanced oil recovery projects.
- Entergy is a participating member of the UtiliTree Carbon Company.

Exelon

- PECO and Exelon have committed $150,000 to TreeVitalize, an aggressive four-year, $8 million partnership to plant more than 20,000 shade trees and restore 1,000 acres of forested riparian buffers in southeastern Pennsylvania.
- Exelon has restored more than 110 acres of natural prairie habitat on buffer lands and rights of way in Illinois since the initiative's beginning in 1994. This effort is helping to sequester CO_2, restore wildlife habitat, prevent runoff, and improve water quality.
- In 2004, restoration work continued on several significant Illinois projects. Exelon partnered with the Forest Preserve District of DuPage County to manage transmission rights of way in conjunction with a larger restoration project.
- Exelon is currently evaluating 10 to 15 additional acres of company rights of way and buffer lands for possible restoration.
- During 2004, Exelon maintained its participation in PowerTree Carbon Company, LLC, an initiative formed in 2003 by 25 U.S. power generators as part of a voluntary industry response to climate change.
- Exelon maintains almost 56,000 miles of overhead electric lines in its distribution system and more than 6,000 miles of transmission rights of way. Its vegetation management program uses safe, reliable, and cost-effective methods, including tree trimming, removal, and herbicide application. These methods follow the standards set by the American National Standards Institute, the Occupational Safety and Health Administration (OSHA) and the International Society of Arboriculture.
- PECO maintains its 12,150 miles of distribution lines on a five-year cycle, ComEd its 43,700 miles on a four-year cycle.

- ComEd is converting sections of transmission rights of way to native grasses and to date has converted 110 acres. Through the Municipal Tree Restoration Program, PECO encourages customers to plant the right tree in the right place to help minimize contact with wires.

Georgia-Pacific

- Georgia-Pacific practices sustainable forest management that results in a large and stable pool of sequestered carbon.

Interface Inc.

- Interface's Cool Carpet™ option allows customers around the world to purchase products with a net-zero climate impact. All of the greenhouse gas emissions associated with modular or broadloom carpet during its entire life cycle are offset through the acquisition of certified emission reduction credits. In the U.S., Canada and Asia-Pacific the credits carry the Climate Neutral Network's "Climate Cool" certification, verifying the accuracy of the life cycle emissions of carpet, and ensuring that the emission reduction credits for carpet sold in these regions are sufficient to achieve a net zero impact on the earth's climate. Interface Europe is working with Climate Care to fund energy reduction projects in South Africa and a major project to re-establish 10,000 hectares of rainforest in Kibale National Park, western Uganda.
- Interface has sponsored the planting of over 52,000 trees since 1997 to offset CO_2 emissions resulting from air travel by its associates through its Trees for Travel program.
- The Interface Cool Fuel™ program enables the company to use corporate gas purchase rebates to offset the CO_2 emissions from business related auto travel in company vehicles.
- The Cool Co$_2$mmute™ program allows Interface associates to offset the CO_2 impact of their personal travel miles by paying for the planting of trees by American Forests to sequester CO_2. Interface splits the cost for the offsets with the associate.

Ontario Power Generation

- Ontario Power Generation plants thousands of trees annually near their operations and service areas, sequestering thousands of tons of CO_2 every year.
- Ontario Power Generation (OPG) and US Gen New England (US Gen) successfully completed a GHG emissions trade in April 2000. US Gen sold OPG 1 million metric tons of CO_2e reductions generated by capturing and destroying methane that would otherwise be emitted from the Johnston Landfill in Rhode Island from 1998 to 2000. OPG has committed to have all of its emissions reduction purchases verified by the Ontario, Canada, Pilot Emissions Trading Project and report them to Canada's Climate Change Voluntary Challenge and Registry Inc., where they are transferred and retired.
- Ontario Power Generation is part of the Greenhouse Emissions Management Consortium, a nonprofit Canadian corporation formed by 10 Canadian Energy Companies that invests in emissions offsets. Among other offsets, the consortium has purchased 6 million metric tons of carbon emission reduction credits from Iowa farmers who use minimum-till and no-till farming practices, cropland retirement, buffer strips, afforestation, reforestation, improved timber management, power generation from biomass, and methane abatement from livestock waste to reduce emissions.

PG&E Corp.

- Pacific Gas and Electric Company has initiated a proposal in California for a new and innovative environmental program that will allow interested customers to contribute toward a cleaner California by offsetting the greenhouse gas emissions associated with their electricity use. This voluntary program is available to most of PG&E's residential and business customers. Through the Climate Protection Program, customers can choose to sign up and pay a small premium on their monthly utility bill which will fund independent environmental projects aimed at removing carbon dioxide from the air.

Rio Tinto

- Rio Tinto's Luzenac America subsidiary purchased green tags from the Bonneville Environmental Foundation to offset 100 percent of the GHG emissions associated with energy used at its Yellowstone Talc mine near Cameron, Montana.

Royal Dutch/Shell

- Royal Dutch/Shell is a member of the CO_2 Capture Research Project, an international effort by seven of the world's leading energy companies.

SC Johnson

- In April 2004, SC Johnson announced a contribution to Conservation International's Conservation Carbon Program to help restore up to 45 acres (18 HA) of degraded forests in the Choco-Manabi corridor of Ecuador. This contribution will support the planting of more than 15 native tree species, and as this regenerated forest grows, it will absorb 5,000 tons of CO_2 over the next 30 years, which will offset the carbon emissions associated with the printing and distribution of the company's annual Public Report.
- SC Johnson, in partnership with the Nature Conservancy, has made a major commitment to help protect the Caatinga, an important bioregion in northeastern Brazil. The project allows for the protection of two sites totaling more than 18,000 acres in the state of Ceara and for the establishment of a local conservation organization to manage the reserves.

TransAlta

- TransAlta is a participating member of the UtiliTree Carbon Company.
- TransAlta, in concert with a coalition of governments and industry, will research commercially viable technology to eliminate CO_2 from coal-burning power plants. The coalition plans to construct and operate a demonstration plant by 2007 to test the technology's technical, environmental, and economic viability.
- TransAlta is part of the Greenhouse Emissions Management Consortium, a nonprofit Canadian corporation formed by ten Canadian Energy Companies that invests in emissions offsets. Among other offsets, the consortium has purchased 6 million metric tons of carbon emission reduction credits from Iowa farmers who use minimum-till and no-till farming practices, cropland retirement, buffer strips, afforestation, reforestation, improved timber management, power generation from biomass, and methane abatement from livestock waste to reduce emissions.

Weyerhaeuser

- At the close of 2004, Weyerhaeuser owned, licensed, or leased 37.9 million acres of forests worldwide. The company uses intensive silvicultural practices on the highly productive forests it owns to achieve the natural biological potential. In other areas it uses less intensive practices to emulate natural forest structure. In both cases, these sustainably managed forests sequester large pools of CO_2 inherent in the trees. Weyerhaeuser invests in afforestation ventures in South America to sustainably sequester additional tons of CO_2 and uses recycled fibers in products to extend the time that CO_2 removed from the atmosphere during the tree-growing stage is stored in products.
- In 2004 Weyerhaeuser improved its process for inventorying GHG emissions and the carbon stored in its forests and products. The company's operations sequestered approximately 26 million metric tons of carbon dioxide equivalents and emitted approximately 7 million tons from the use of fossil fuels and other activities. This effectively sequestered 19 million metric tons of carbon dioxide equivalent or 0.5 metric tons of carbon equivalents per ton of production, an improvement of approximately 18 percent over 2003.

Wisconsin Energy Corp.

- Wisconsin Energy is a partner in the Rio Bravo Carbon Sequestration Project to protect 65,000 acres of endangered rainforest in Belize. The project combines land acquisition and sustainable forestry and is expected to sequester approximately 2.4 million metric tons of carbon over 40 years.
- Wisconsin Energy is a participating member of the UtiliTree Carbon Company.
- Wisconsin Energy funds EPRI research to assess the potential of existing options for capturing and sequestering CO_2 emissions; evaluate methods for CO_2 capture and sequestration at the point of electricity generation; and investigate enhanced terrestrial or oceanic processes that remove and store atmospheric CO_2.

Emissions Trading, Joint Implementation (JI) and Clean Development Mechanism (CDM) Solutions

ABB

- ABB has built power plants in Costa Rica through a mutually beneficial climate improvement project as part of a Norwegian consortium. The project could avoid an estimated 4 million tons of CO_2 emissions over a 20-year period. Any CO_2 credits earned by Costa Rica could then be sold to Norwegian companies or other buyers.
- ABB participates in an Activities Implemented Jointly (AIJ) reforestation and forest conservation project in Costa Rica that will sequester an estimated net 230,800 metric tons of carbon over its 25-year lifetime.
- ABB funds and manages the China Energy Technology Program, a two-year project evaluating the costs and environmental impacts of various technological options to provide sustainable electricity generation in the Shandong Province of China.

American Electric Power

- AEP is part of a collaborative GHG mitigation pilot project with the Government of Bolivia, the Nature Conservancy, and the Bolivian Friends of Nature Foundation. The Noel Kempff Mercado Climate Action Project will protect nearly 4 million acres of threatened forest and offset 5 to 7 million tons of carbon over 30 years.
- AEP is a partner in the Guaraqueçaba Climate Action Project, which seeks to restore and protect nearly 20,000 acres of partially degraded and/or deforested land in the tropical Atlantic rainforest of Brazil. The Project is expected to offset approximately 1 million metric tons of carbon over 40 years.
- AEP is a member of e8, a consortium of nine of the world's leading electric companies from G8 countries. e8 promotes sustainable energy projects through a "learn by doing" approach on electricity-related issues in developing countries with host countries, UN agencies, NGOs, and local energy providers. e8 also works to develop human capacity building—an example being its Micro-Solar Distance Learning Program, which focuses on electrification for information and telecommunications needs using photovoltaics. AEP serves as the U.S. delegate to the e8 and has undertaken the project leadership for an e8 effort to install wind turbines on environmentally sensitive San Cristobal Island in the Galapagos.
- AEP has led the creation of Global 3E, a charitable organization designed to attract $60 million in private contributions from the philanthropic sector in two years. Global 3E will provide zero interest loans for electricity-related humanitarian projects in developing nations.
- AEP is a founding and active member of the Chicago Climate Exchange (CCX). Through CCX, AEP has made legally binding commitments to reduce its GHG emissions by 4 percent below the average of its 1998 to 2001 baseline by 2006.

Baxter International

- Baxter has joined CCX, a group of organizations including the City of Chicago and Mexico City, to pilot a voluntary carbon trading system focused on reducing absolute aggregated group greenhouse gas emissions a certain amount over a period of time.

BP

- BP believes that the use of flexible market mechanisms, such as emissions trading and the CDM, provide a cost-effective means of reducing greenhouse gas emissions. BP operated an internal emissions trading system between 1999 and 2001 that helped reduce operational emissions by 10 percent. The system covered all BP operations across the globe and provided a number of insights and learning.
- BP's UK Upstream and Petrochemicals assets are now part of the UK Government's emissions trading scheme (ETS). BP carried out the first trades in the UK ETS and has also helped customers trade in the market. BP is currently applying the evolving CDM rules and procedures to a real BP solar project in Brazil, with the intention of registering the project with the CDM Executive Board. BP is currently piloting CDM projects for a range of technologies and believes clear accounting principles need to be created and internationally agreed upon, for the use of CER's and other types of GHG emission reduction credits, to realize value from lower carbon technologies, and for compliance use in meeting mandated and voluntary GHG emission caps.

Cinergy Corp. (now Duke Energy)

- Cinergy Corp. is working with other industries and organizations to pilot emissions trading systems, and through its subsidiary company United States Energy Bio-gas has completed the trading of carbon equivalent offsets with a Canadian company.
- Cinergy Corporation is a partner in the Rio Bravo Carbon Sequestration project to protect 65,000 acres of endangered rainforest in Belize. The project combines land acquisition and a sustainable forestry program and is expected to sequester approximately 2.4 million metric tons of carbon over 40 years. Cinergy, the Nature Conservancy, and the Belize Government entered into an agreement to transfer the carbon offsets from the project to Cinergy Corporation. The agreement was reviewed and approved by U.S. Initiative on Joint Implementation (USIJI).

DTE Energy

- DTE Energy, along with other partners, is involved in the Rio Bravo Carbon Sequestration Project to protect 65,000 acres of endangered rainforest in Belize. The project combines land acquisition and sustainable forestry and is expected to sequester approximately 2.4 million metric tons of carbon over 40 years.

DuPont

- DuPont has been active in working with others to pilot emissions trading systems and has concluded a number of trades through the use of bi-lateral agreements and on the emerging carbon exchanges. DuPont is a member of the Chicago Climate Exchange and the International Emissions Trading Association. In the winter of 2002, DuPont donated 120,000 tons of CO_2-equivalent emissions credits to the Salt Lake City Organizing Committee. This allowed the Winter Olympics to offset their emissions and be declared "climate neutral."

Entergy

- Entergy and Elsam, the largest Danish electricity supplier, executed an international trade in CO_2 allowances under the Danish climate change program. Under the transaction, Entergy purchased 10,000 Danish allowances from Elsam and will remove the allowances from the market, eliminating 10,000 metric tons of CO_2 emissions.

Interface Inc.

- Interface Inc. joined the Chicago Climate Exchange (CCX) in November 2004. They were the first and only company in the commercial interiors industry to do so. Through CCX, Interface is actively quantifying, tracking, and reporting the GHG emissions associated with the manufacture of its products (carpet and fabric) in the United States and Canada.
- Interface has committed to decrease its absolute GHG emissions by 1 percent per year through 2006, using the period from 1998 to 2001 as a baseline, and Interface's North American business units have committed to support emerging markets for GHG emissions trading in North America.
- Interface purchases a range of carbon offsets to support its Cool Carpet™ program, enabling its customers to purchase "climate neutral" flooring products. The emission reduction credits associated with Interface's Cool Carpet products offset the full product life cycle—raw material extraction to end-of-life.

Ontario Power Generation

- Ontario Power Generation (OPG) and PG&E's subsidiary US Gen New England (US Gen) successfully completed a GHG emissions trade in April 2000. US Gen sold OPG 1 million metric tons of CO_2 equivalent emissions reductions generated by capturing and destroying methane that would otherwise be emitted from the Johnston Landfill in Rhode Island from 1998–2000. OPG has committed to have all of its emissions reduction purchases, such as this one, verified by the Ontario, Canada, Pilot Emissions Trading Project (PERT) and report them to Canada's Climate Change Voluntary Challenge and Registry (VCR) Inc., where they are transferred and retired.
- Ontario Power Generation has registered 1.8 million metric tons of CO_2 emission reduction credits produced by its in-house energy efficiency program. OPG also purchased emission reduction credits equal to approximately 10 million metric tons of CO_2 to meet its voluntary 26 million metric ton emissions target for the year 2000. The purchased emissions resulted from a variety of offset activities, including carbon sequestration in forests and agricultural lands, power generation from biomass, livestock waste methane abatement, and the extraction of CO_2 from natural gas.
- OPG is part of the e8, a consortium of nine of the world's leading electric companies from G8 countries.
- Ontario Power Generation met its year 2000 net GHG emission target by offsetting almost 33 percent of its greenhouse gas emissions with emission reduction credits. Of these, 2.3 million metric tons were generated from internal energy efficiency improvements. The remaining credits were purchased from sources in North America and internationally. OPG has also transferred the 12.6 million metric tons CO_2 equivalent emission reduction credits to Canada's Climate Change Voluntary Challenge and Registry Inc. for retirement.

Royal Dutch/Shell

- Royal Dutch/Shell Group developed and used a pilot internal emissions trading system (STEPS) to gain experience and understanding in the use of and structure for emissions trading. The system, which ran from 2000 to 2002, allowed trading between a number of Group entities located in Annex 1 countries. The system covered over 33 million metric tons of CO_2e from over 22 separate sites, accounting for almost two-thirds of Shell's developed country emissions or over one-third of its global emissions.
- Shell has shifted emphasis from internal mechanisms to real external instruments and has established an Environmental Products Trading Business (EPTB). The Shell Group has entered the UK Emissions Trading System, and as a result, key Shell UK upstream production facilities now have a GHG emissions cap. Shell Trading, with Nuon, executed the first trade in EU CO_2 allowances in February 2003.

TransAlta

- In August of 2004, TransAlta announced the purchase of 1.75 million tonnes of GHG candidate Certified Emission Reductions (CERs) from the Chilean agricultural company Agrosuper. The purchase requires the registration of the project with the CDM Executive Board. This agreement represents the first Canadian purchase of CERs under the Kyoto Protocol.
- TransAlta develops and trades for approximately 4 million tonnes of CO_2 equivalent per year in offset projects, with 80 million tons currently under contract. Offset projects include gas recovery, energy efficiency, ruminant methane, landfill and coal mine gas to electricity, forestry, and soil sequestration, among others. In a recent upgrade of its U.S. operations, TransAlta reduced its CO_2 emissions by an amount equal to the annual emissions of 27,800 cars and sold the resulting credits to a U.S. integrated oil company.
- TransAlta contributes to the development of a greenhouse gas emissions reduction market by engaging in selling fractions of its portfolio.

Wisconsin Energy Corp.

- Wisconsin Energy participates in a project that involves fuel-switching (coal to natural gas), cogeneration, and efficiency improvements to a district heating plant in the City of Decin, Czech Republic. This project has improved local air quality, reduced greenhouse gas and other emissions, and provided educational opportunities and experience for other communities interested in improving air quality.

Whirpool

- Whirlpool, in cooperation with government agencies, utilities, NGOs, and manufacturers, has a program to encourage the early retirement of inefficient appliances in Brazil. This program can potentially avoid more than 3 million tons of CO_2 emissions each year.

Appendix B

Glossary

1605(b): Under the Energy Policy Act (EPAct) of 1992, Section 1605(b) program companies are encouraged by the Department of Energy to voluntarily report activities undertaken to reduce GHG emissions or to sequester carbon. Companies may want to report these activities to achieve recognition of achievements (from both regulators and stakeholders), inform the public debate on climate change, or to participate in educational exchanges.

Carbon Dioxide Equivalent (CO_2e): A metric used to compare the emissions from various greenhouse gases based upon their global warming potential (GWP). Carbon dioxide equivalents are commonly expressed as "million metric tons of carbon dioxide equivalents (MMTCDE)." They provide a universal standard of measurement against which the impacts of releasing (or avoiding the release of) different greenhouse gases can be evaluated. Every greenhouse gas has a Global Warming Potential (GWP), a measurement of the impact that particular gas has on "radiative forcing;" that is, the additional heat/energy which is retained in the Earth's ecosystem through the addition of this gas to the atmosphere. The GWP of a given gas describes its effect on climate change relative to a similar amount of carbon dioxide and is divided into a three-part "time horizon" of twenty, one hundred, and five hundred years. As the base unit, carbon dioxide numeric is 1.0 across each time horizon. This allows the greenhouse gases regulated under the Kyoto Protocol to be converted to the common unit of CO_2e. Global Warming potentials for the greenhouse gases regulated under the Kyoto Protocol under a 100 year timeframe are as follows: Carbon dioxide (CO_2) has a GWP of 1; Methane (CH_4) has a GWP of 23; Nitrous oxide (N_2O) has a GWP of 296; Halocarbons (HFC) has a GWP range from 140 for HFC-152a to 11,700 for HFC-23; Perfluorocarbons have a GWP range from 6,500 to 9,200; Sulfur Hexafluoride (SF_6) has a GWP of 23,900.[158]

Certified Emissions Reduction (CER): Reductions of greenhouse gases achieved by a Clean Development Mechanism (CDM) project. A CER can be sold or counted toward Annex I countries' emissions commitments. Reductions must be additional to any that would otherwise occur.

Chlorofluorocarbons (CFCs): Compounds consisting of chlorine, fluorine, and carbon. CFCs are very stable in the troposphere, however are broken down by strong ultraviolet light in the stratosphere to release chlorine atoms that deplete the ozone layer. CFCs are commonly used as refrigerants, solvents and foam blowing agents. International phase-out programs of these chemicals are in place, most importantly the 1987 Montreal Protocol and its subsequent amendments. CFCs are also considered to be greenhouse gases and are targeted for reduction under the 1997 Kyoto Protocol.

Clean Development Mechanism (CDM): One of the three market mechanisms established by the Kyoto Protocol. The CDM is designed to promote sustainable development in developing countries and assist Annex I Parties in meeting their greenhouse gas emissions reduction commitments. It enables industrialized countries to invest in emission reduction projects in developing countries and to receive credits for reductions achieved.

Direct Emissions: Emissions from sources owned by the reporter.

Emissions Trading: A market mechanism that allows emitters (countries, companies or facilities) to buy emissions ("permits" or "credits") from or sell emissions to other emitters. Emissions trading is expected to bring down the costs of meeting emission targets by allowing those who can achieve reductions less expensively to sell excess reductions (e.g. reductions in excess of those required under some regulation) to those for whom achieving reductions is more costly.

Geologic Sequestration: Injecting captured CO_2, under pressure into stable geologic formations where it is expected to remain indefinitely.

Global Warming Potential (GWP): See explanation under CO_2 equivalent (CO_2e).[159]

Greenhouse Gases: There are six focal greenhouse gases. Greenhouse gases that are both naturally occurring and manmade include *carbon dioxide* (CO_2), *methane* (CH_4), and *nitrous oxide* (N_2O). Greenhouse gases that are not naturally occurring include *hydrofluorocarbons* (HFCs), *perfluorocarbons* (PFCs), and *sulfur hexafluoride* (SF_6).

Hydrochlorofluorocarbons (HCFCs): HCFCs are synthetic industrial gases made up of hydrogen, chlorine, fluorine and carbon. They are being used as commercial substitutes for chlorofluorocarbons (CFCs) primarily for refrigeration but also as blowing agents for insulating plastic foams, fire extinguisants, and solvents. There are no natural sources of HCFCs. These compounds deplete stratospheric ozone, although much less than CFCs. Production and consumption of these gases are controlled under the Montreal Protocol.

Hydrofluorocarbons (HFCs): HFCs are used as a replacement for CFCs in a variety of industrial processes, including semiconductor manufacture (plasma etching and cleaning tool chambers), refrigeration and fire protection and have been used as a replacement for CFCs. The atmospheric lifetime of HFCs ranges from about 1.5 years for HFC-152a to over 250 years for HFC-23. HFCs are among the six greenhouse gases to be curbed under the Kyoto Protocol.

IGCC: Integrated Gasification Combined Cycle plants gasify coal, biomass, or petroleum waste products (typically from refining processes) without burning the feedstock. The gas is then combusted in a gas turbine, and waste heat is used to create steam to drive a steam turbine. Sulfur dioxide and other trace impurities are removed prior to combusting the gas. The process uses less water and produces approximately 50 percent less solid waste than conventional coal-fired plants (which combust pulverized coal to create steam) and produces a pure carbon dioxide stream that can be separated and captured with a lower energy penalty and at lower incremental costs than in the case of pulverized coal plants. Another benefit is the potential to remove mercury at lower costs than in conventional coal-fired plants.

Indirect Emissions: Indirect emissions are defined as emissions from sources other than that owned by the reporter, but caused by actions on the part of the reporter. The predominant source of indirect emissions is the purchase or sale of electricity. Another source of indirect emissions might include emissions caused by product use (i.e. the calculated emissions of the fleet of GM vehicles in operation in the United States or of the operation of Whirlpool washers and dryers in the United States). There are clear problems with these measures. For example, there is a real risk of double counting as both a utility and the entity that purchases the electricity each counts the emissions for the same kilowatt. The key question becomes, who "owns" the emissions arising from power generated on behalf of others.[160]

Kyoto Protocol: An international agreement adopted in December 1997 in Kyoto, Japan. The Protocol sets binding emission targets for developed countries that would reduce their emissions on average 5.2 percent below 1990 levels.

Make-Rate: A term to describe the weight ratio of HFC-23 byproduct to HCFC production expressed as a percentage.

McCain-Lieberman Climate Stewardship Act: A bipartisan national plan for action to begin solving the problem of global warming. The Act gives power plants, oil companies and factories until 2010 to collectively reduce their greenhouse emissions to the levels they emitted in 2000. The Act calls for the creation of an emissions trading system to help companies meet these requirements. The Act also allows companies to meet a portion of their emissions goal by paying farmers to use conservation methods to increase the amount of carbon stored in their soil.

Nitrous Oxide (N_2O): N_2O is among the six greenhouse gases to be curbed under the Kyoto Protocol. N_2O is produced by natural processes, but there are also substantial emissions from human activities such as agriculture, industrial production of nitric acid and adipic acid and fossil fuel combustion. The atmospheric lifetime of N_2O is over 100 years, and its GWP is 310.

Off-System Reductions: GHG emission reductions that are achieved outside of the company's operations, such as reforestations projects (biological sequestration) or energy conservation projects undertaken with customers.

On-System Reductions: GHG emission reductions that are achieved within the company's operations, such as heat rate improvement projects at electricity generation stations, renewable energy demonstration projects or implementation of hybrid vehicles.

Safety Valve: A price cap within the cap-and-trade program whereby participants can purchase allowances from the government at the safety valve price if market prices exceed the safety valve. This would lower the risks of economic shocks created by unexpectedly high allowance prices, while lowering the risks of such a program being rolled back if high prices emerged (such as happened in the California RECLAIM market, where NO_x prices exceeded $40,000/ton, causing the program to be shut down). Such a program is often referred to as a "hybrid," combining elements of a cap-and-trade program with those of an emissions tax.

Sequestration: Opportunities to remove atmospheric CO_2, either through biological processes (e.g. plants and trees), or geological processes through storage of CO_2 in underground reservoirs.

Notes

1. Margolick, M. and D. Russell. 2001. *Corporate Greenhouse Gas Reduction Targets.* (Washington DC: Pew Center on Global Climate Change).

2. Wellington, F. and A. Sauer. 2005. *Framing Climate Risk in Portfolio Management.* (Boston, MA: Ceres).

3. Karmali, A. 2006. "Best Practice in Strategies for Managing Carbon." in K. Tang (ed.) *The Finance of Climate Change: A Guide for Governments, Corporations and Investors.* (London: Risk Books): 259–270.

4. Austin, D. and A. Sauer. 2002. *Changing Oil: Emerging Environmental Risks and Shareholder Value in the Oil and Gas Industry.* (Washington DC: World Resources Institute).

5. DeCicco, J., F. Fung and F. An. 2005. *Automakers' Corporate Carbon Burdens: Update for 1990–2003.* (New York: Environmental Defense).

6. Cogan, D. 2006. *Corporate Governance and Climate Change: Making the Connection.* (Boston, MA: Ceres Inc.).

7. Hoffman, A. 2001. *From Heresy to Dogma: An Institutional History of Corporate Environmentalism.* (Stanford, CA: Stanford University Press).

8. United States Senate Committee on Energy and Natural Resources. 2006. "Climate Change Conference." April 4, 110th Congress, 2nd session. Washington, DC: Government Printing Office. http://frwebgate.access.gpo.gov/cgi-bin/getdoc.cgi?dbname=109_senate_hearings&docid=f:28095.pdf, viewed 3/7/07.

9. Fialka, J. 2007. "Democrats Push Climate Change to Front Burner." *Wall Street Journal,* January 18: A1.

10. For more on the U.S. Climate Action Partnership, go to: http://www.us-cap.org/, viewed 1/29/07.

11. Semans, T. 2007. "Climate Change as a Driver of U.S. Market Behavior." [Weblog entry] Turning the Ship. *Greenbiz.com.* Feb. 8. http://blog.turningtheship.com/?p=11, viewed 3/7/07.

12. *New York Times.* 2006. "Investors Seek Climate Change Information." *New York Times,* June 15: C8.

13. Epstein, P. and E. Mills. 2005. *Climate Change Futures: Health, Ecological and Economic Dimensions.* (Cambridge, MA: Center for Health and the Global Environment, Harvard Medical School).

14. Sorkin, A. 2007. "A Buyout Deal That Has Many Shades of Green." *New York Times.* February 26.

15. Smith, J. 2005. "The Implications of the Kyoto Protocol and the Global Warming Debate for Business Transactions." *NYU Journal of Law & Business,* 1(2): 511:550; Ewing, K., J. Hutt and E. Petersen. 2004. "Corporate Environmental Disclosures: Old Complaints, New Expectations." *Business Law International,* 5(3): 459-515.

16. Semans, T. and T. Juliani. 2006. "Succeeding in a Carbon-Constrained World." *Corporate Strategy Today,* July.

17. See: Interfaith Center on Corporate Responsibility, http://www.iccr.org/, viewed 3/3/06.

18. Carbon Trust. 2004. *Brand Value at Risk from Climate Change.* (London: Carbon Trust).

19. Pacala, S. and R. Socolow. 2004. "Stabilization Wedges: Solving the Climate Problem for the Next 50 Years with Current Technologies." *Science,* 305(5686): 968-972.

20. Makower, J., R. Pernick and C. Wilder. 2007. *Clean Energy Trends 2007.* (San Francisco: Clean Edge Inc.) http://www.cleanedge.com/reports/Trends2007.pdf, viewed 3/7/07.

21. Chea, T. 2006. "Doerr Firm Invests in 'Green Technology'." *USA Today online,* April 10. http://www.usatoday.com/tech/news/2006-04-10-green-venture-capitalist_x.htm, viewed 6/22/06.

22. Notably, in a 2003 scenario planning exercise involving top global industry, academic, and government experts published by the Pew Center on Global Climate Change, the worst-case energy price inputs projected out to 2035 were all surpassed as of the writing of this report. See Mintzer, I., J. Leonard and P. Schwartz. 2003. *U.S. Energy Scenarios for the 21st Century.* (Washington, DC, Pew Center on Global Climate Change).

23. See the Swiss Re case study in this report.

24. The industry has noted that the magnitude of losses is not only due to more disasters but also to more and higher priced property being in disaster areas (i.e., million-dollar beachfront properties).

25. Ceres. 2006. *Managing the Risks and Opportunities of Climate Change: A Practical Toolkit for Corporate Leaders.* (Boston, MA: Ceres Inc).

26. Carey, J. 2006. "Business on a Warmer Climate." *Business Week,* July 17: 26-29.

27. Lemonick, M. 2005. "Has the Meltdown Begun?" *Time,* February 27: 58-59.

28. Eilperin, J. 2006. "Debate on Climate Change Shifts to Irreparable Harm." *Washington Post,* January 29: A01.

29. Deutsch, C. 2006. "Study Says U.S. Companies Lag on Global Warming." *New York Times,* March 22: C3.

30. Birchall, J. 2005. "Wal-Mart Sets Out Stall for a Greener Future." *The Financial Times,* October 26: 27.

31. Bennett, O. 2006. "Healthy, Wealthy And Wise Inc. 'Sustainability' Used To Be Just For Hippies. In America, It's Now Big Business." *The Daily Telegraph,* October 25: 8.

32. Crane, D. 2004. "Canada Needs to Develop a Clear Plan on Kyoto." *The Toronto Star,* September 18: D2.

33. For more on the Stern Report, go to: http://www.hm-treasury.gov.uk/independent_reviews/stern_review_economics_climate_change/stern_review_report.cfm, viewed 1/12/07.

34. BELC survey respondents are AEP, Air Products, Alcan, Alcoa, Baxter, BP, Cinergy, DTE Energy, DuPont, Entergy, Exelon, Georgia-Pacific, Holcim, IBM, Intel, Interface, Maytag, PG&E, Rio Tinto, Rohm and Haas, SC Johnson, Shell, Sunoco, TransAlta, United Technologies, Whirlpool, and Wisconsin Energy.

35. Non-BELC survey respondents are Advanced Micro Devices, Calpine, Fairchild Semiconductors, and Staples.

36. BELC case studies are Alcoa, Cinergy, DuPont, The Shell Group and Whirlpool.

37. Non-BELC case study is Swiss Re.

38. For more information on the Clean Development Mechanism, go to: http://cdm.unfccc.int/, viewed 3/3/06.

39. Buaron, R. 1981. "New-Game Strategies." *The McKinsey Quarterly*: 36.

40. Moran, M., A. Cohen, N. Swem, and K. Shaustyuk. 2005. "The Growing Interest in Environmental Issues is Important to Both Socially Responsible and Fundamental Investors." *Portfolio Strategy.* Goldman Sachs, August 26: 5.

41. SAP refers to Systemanalyse und Programmentwicklung ("Systems Analysis and Program Development"), standard enterprise software which integrates all business processes. It was developed in 1972 by five former IBM employees.

42. For more information on the IPCC, go to: http://www.ipcc.ch/, viewed 3/3/06.

43. The terms "direct" and "indirect" as used in this document should not be confused with their use in national GHG inventories where "direct" refers to the six Kyoto gases and "indirect" refers to certain GHG precursors.

44. For more information on the WRI/WBCSD Greenhouse Gas Protocol Corporate Accounting and Reporting Standard, go to: http://www.ghgprotocol.org, viewed 3/3/06.

45. Some companies use both indexed and absolute measures.

46. Absent from these top four is the indexed metric included in the President's Global Climate Change Initiative, which sets goals in terms of the GHG intensity of the economy (in tons of emissions per dollar of GDP). The National Commission on Energy Policy (NCEP) also recommended this measure. While the emissions-per-dollar-GDP metric was designed for national policy and not for individual companies, its absence is notable because companies often develop internal protocols to match external reporting and compliance structures. National Commission on Energy Policy. 2004. *Ending the Energy Stalemate: A Bipartisan Strategy to Meet America's Energy Challenges.* (Washington DC: National Commission on Energy Policy): 21.

47. It should be noted that all utilities are required to install CEMs for units that generate 25 MW or larger under the Clean Air Act.

48. DuPont. 2006. *Press Release: DuPont and BP Announce Partnership to Develop Advanced Biofuels.* (Wilmington: DuPont).

49. In May, 2006, Entergy announced its second GHG reduction commitment target. It made a voluntary commitment to reduce greenhouse gas emissions from its operating plants and stabilize those emissions at a level 20 percent below year 2000 from 2006–2010.

50. Prepared by Truman Semans of the Pew Center, drawing on consulting engagements and extensive discussions with the Pew Center's BELC from March 2005 to May 2006.

51. This financial savings figure is calculated as the costs avoided through energy reductions achieved by improving yields and creating less energy intensive product portfolios versus the business as usual scenario.

52. Little, A. 2006. "Don't Discount Him: An Interview with Wal-Mart CEO H. Lee Scott." *Grist Magazine,* April 12. http://www.grist.org/news/maindish/2006/04/12/griscom-little/index.html?source=daily, viewed 4/20/06.

53. Op. cit., Carey, J. 2006.

54. Ibid., Carey, J. 2006.

55. Exelon Corp. 2006. *Carbon Disclosure Project, Fourth Edition, Greenhouse Gas Questionnaire Response.*

56. In assessing existing or proposed public policies, analysts often need shadow price estimates for one or two key items—for example, the value of a human life, the cost of various types of injuries, the cost of a robbery or the value of an hour of travel time saved, or the cost of carbon emissions.

57. The Chicago Climate Exchange (CCX) is North America's only, and the world's first, greenhouse gas (GHG) emission registry, reduction and trading system for all six greenhouse gases (GHGs). CCX is a self-regulatory, rules based exchange designed and governed by CCX Members. Members make a voluntary but legally binding commitment to reduce GHG emissions. For more information, go to http://www.chicagoclimatex.com.

58. Malone, T. 2004. "Bringing the Market Inside." *Harvard Business Review,* April: 107-114.

59. Morgheim, J. 2000. *Testimony before the Senate Commerce, Science and Transportation Committee,* September 21.

60. Baxter International. 2006. *Carbon Disclosure Project, Fourth Edition, Greenhouse Gas Questionnaire Response.*

61. Kerr, S. 1995. "On the Folly of Rewarding A While Hoping for B." *Academy of Management Executive,* 9(1): 7-16.

62. For more on the Smart Trips program, see: http://www.smarttrips.org/, viewed 3/3/06.

63. For more on the One-Ton Challenge program, see: http://www.climatechange.gc.ca/onetonne/english/index.asp?pid=179, viewed 3/3/06.

64. See: http://www.hybridcars.com/corporate-incentives.html, viewed 3/3/06.

65. Jick, T. 1993. *Managing Change.* (Boston, MA: Irwin McGraw-Hill).

66. Op. cit., Hoffman, A. 2001.

67. Hoffman, A. 1996. "Environmental Management Withers Away." *Tomorrow,* 6(2): 60-61.

68. Howard-Grenville, J. and A. Hoffman. 2003. "The Importance of Cultural Framing to the Success of Social Initiatives In Business." *Academy of Management Executive,* 17(2): 70-84.

69. Sullivan, N. and R. Schiafo. 2005. "Talking Green, Acting Dirty." *New York Times,* June 12: 23.

70. This case study was written by Andre de Fontaine (Markets & Business Strategy Intern, Pew Center on Global Climate Change) based on interviews with Andrew Ruben (Vice President for Corporate Strategies and Sustainability, Wal-Mart) on Aug. 10 and 16, 2006 and Jib Ellison (Managing Partner, Blu Skye Sustainability Consulting) on Aug. 18, 2006.

71. Scott, L. 2005. *21st Century Leadership.* Speech to company employees, October 24 (Bentonville, AR: Wal-Mart).

72. Op. cit., National Commission on Energy Policy. 2004: 21.

73. See: Lyon, T. and J. Maxwell. 2004. *Corporate Environmentalism and Public Policy.* (Cambridge: Cambridge University Press); Lyon, T. and J. Maxwell. 2003. "Mandatory and Voluntary Approaches to Mitigating Climate Change." In A. Baranzini and P. Thalmann (eds.), *Voluntary Agreements in Climate Policies.* (London: Edward Elgar Press).

74. Bodansky, D., S. Chou and C. Jorge-Tresolini. 2004. *International Climate Efforts Beyond 2012: A Survey of Approaches.* (Arlington, VA: Pew Center on Global Climate Change).

75. Deloitte Research. 2005. *Which Way to Value? The U.S. Power and Utility Sector, 2005-2010.* (Washington DC: Deloitte Development LLC).

76. For more on the Coal21 Fund, see: www.coal21.com, viewed 3/3/06.

77. Tully, S. 2005. "Commercial Contributions to the Climate Change Regime: Who's Regulating Whom?" *Sustainable Development Law and Policy,* 5(2): 14-27.

78. Duke Energy Corp. 2006. *Executive Summary of Responses by Duke Energy Corp. to Questions Posed in a February 2006 White Paper Entitled: Design Elements of a Mandatory Market-Based Greenhouse Gas Regulatory System.* (Charlotte, NC: Duke Energy Corp.).

79. Hockenstein, J., R. Stavins and B. Whitehead. 1997. "Crafting the Next Generation of Market-Based Environmental Tools." *Environment,* 39(4): 12-20, 30-33.

80. The federal securities laws require publicly traded companies to disclose information on an ongoing basis. For example, domestic issuers must submit annual reports on Form 10-K, quarterly reports on Form 10-Q, and current reports

on Form 8-K for a number of specified events and must comply with a variety of other disclosure requirements. The annual report on Form 10-K provides a comprehensive overview of the company's business and financial condition and includes audited financial statements. http://www.sec.gov/answers/form10k.htm, viewed 9/25/06.

81. The goal of the CDP questionnaire is to understand the possible material impacts on the value of investments driven by climate change related to: taxation and regulation; changes in weather patterns; technological innovations and shifts in consumer attitude and demand. Specifically, the questionnaire asks companies to explain: the quantity of emissions they produce; their goals and plan for reducing them; the company's total cost of energy consumption; how climate change might represent commercial risks and/or opportunities to the company; the financial and strategic impacts of existing regulation of GHG emissions; how operations might be affected by extreme weather events; what technologies, products, processes or services the company has developed, or is developing, in response to climate change; who at the board level has specific responsibility for climate change related issues and who manages your company's climate change strategies; the strategy for communicating the risks and opportunities from GHG emissions and climate change through an annual report and other communications channels; and the firm's strategy for, and expected cost/profit from trading in the E.U. Emissions Trading Scheme, CDM/JI projects and other trading systems. One gap in the CDP questionnaire to date is a requirement that firms disclose how they engage on public policy engagement related to climate and the policy positions they advocate. Thus, a firm that CDP scores well on disclosure and operational efforts to reduce emissions may also be opposing progress toward legislation to cap GHGs.

82. See: http://www.cdproject.net/, viewed 3/3/06.

83. Op. cit., Smith, J. 2005.

84. *The Economist*. 2006. "Companies and Climate Change: Can Business Be Cool?" *The Economist*, June 10, 379(8481): 70.

85. Wellington, F. 2005. *Framing Climate Change Risk in Portfolio Management*. (Washington DC: World Resources Institute): 3.

86. Ibid., Wellington, F. 2005.

87. For more on Goldman Sachs' environmental policy framework, go to: http://www.gs.com/our_firm/our_culture/social_responsibility/environmental_policy_framework/index.html, viewed 3/3/06.

88. Suozzo, P. 2006. "Investing in Solutions to Climate Change" *Citigroup Research*, June 12: 7.

89. Deutsch, C. 2006. "Wall St. Develops the Tools To Invest in Climate Change." *New York Times*, May 24: C3.

90. Op. cit., Deloitte Research. 2005.

91. Another problem is that there is a lack of harmonization of the standards. For example, as of June, 2006, 31 states used the AASHTO specification standards and 16 used the ASTM standards. Only 4 states used both (source: Portland Cement Association).

92. O'Driscoll, M. 2006. "Duke's Rogers Will Follow EEI Line on CO_2 Reductions." *Greenwire*, http://www.eenews.net/gw/, viewed 6/21/06.

93. Aston, A. and Helm, B. 2005. "The Race Against Climate Change." *Business Week*, December 12: 59-66, 132.

94. Top Scores for BELC members: DuPont (Chemicals), AEP, Cinergy, Entergy, Exelon and PG&E (Electric Power), Toyota (Autos), GE and ABB (Industrial Equipment), Alcan and Alcoa (Metals and Mining), BP and Shell (Oil and Gas) and Rio Tinto (Coal).

95. Op. cit., Cogan, D. 2006.

96. GHG controls will create new markets in emissions allowances, capital, and technologies that are efficient in their use of energy or creation of carbon.

97. Hoffman, A. 2005. "Climate Change Strategy: The Business Logic Behind Voluntary Greenhouse Gas Reductions." *California Management Review,* 47(3): 21-46.

98. Cinergy 2005. *About Cinergy, Cinergy at a Glance*, http://www.cinergy.com/About_Cinergy_Corp/Corporate_Overview/default_at_a_glance.asp, viewed 10/9/05.

99. Ibid., Cinergy 2005. *About Cinergy, Cinergy at a Glance.*

100. Based upon the Department of Energy (Energy Information Administration) figure of 1.016 billion tons of coal consumed for electric power in 2004.

101. Based upon Department of Energy (Energy Information Administration) 2004 data, combustion of coal accounted for approximately 85 percent of CO_2 emissions from the electric power sector in the United States in 2004. The electric power sector accounted for approximately 40 percent of CO_2 emissions in the United States. Total CO_2e emissions in the United States were approximately 7.6 billion tons in 2004.

102. Cinergy. 2004. *Air Issues Report to Stakeholders.* (Cincinnati, Ohio: Cinergy Corp.): 26.

103. David, J. and H. Herzog. 2002. *The Cost of Carbon Capture.* (Washington DC: U.S. DOE, National Energy Technology Laboratory) http://www.netl.doe.gov/publications/proceedings/01/carbon_seq_wksp/David-Herzog.pdf, viewed 2/7/06.

104. Op. cit., Cinergy. 2004. *Air Issues Report:* 29.

105. Ibid., Cinergy. 2004, *Air Issues Report*: 2.

106. Ibid., Cinergy. 2004, *Air Issues Report:* 2.

107. Ibid., Cinergy. 2004, *Air Issues Report:* 26.

108. Ibid., Cinergy. 2004, *Air Issues Report:* 26.

109. Cinergy. 2004. *Annual Report.* (Cincinnati, OH: Cinergy Corp.): pp. 14-15.

110. Op. cit., Cinergy. 2004, *Air Issues Report.*

111. The full scale of Duke Energy is changing as it has announced the sale of Duke Energy North America (DENA) and its 6,200 MW of generating capacity for $1.54 billion. Duke Energy 2006. "Duke Energy Announces Sale of DENA Power Generation Assets to LS Power." http://www.duke-energy.com/news/releases/2006/jan/2006010901.asp, viewed 2/1/06.

112. UNFI. 2002. *CEO Briefing on Climate Change.* http://unepfi.org, viewed 1/29/06.

113. Swiss Re. 2004. *Sustainability Report 2004*, http://www.swissre.com, viewed 1/29/06.

114. Swiss Re. 2004. *Climate Change Matters to All of Us: The Great Warming*, http://www.swissre.com/, viewed 1/24/06.

115. Schmidheiny, S. and F. Zorraquin. 1996. *Financing Change: the Financial Community, Eco-Efficiency, and Sustainable Development.* (Cambridge, MA: MIT Press).

116. Aston, A. and Helm, B. 2005. "Financial Services Leaders." *Business Week*, December 12.

117. Reinsurance is a contract by which an insurer is insured wholly or in part against the risk he has incurred in insuring somebody else.

118. Swiss Re. 1994. *Global Warming, Elements of Risk.* (Zurich: Swiss Re).

119. In December 2004, HSBC made a commitment to become the world's first major bank to achieve carbon neutrality. The target was achieved in 2006.

120. Swiss Re also works to meet MINERGIE® standards, the European equivalent of LEED®, in all its new buildings in Switzerland, and in refurbishing projects whenever financially, technically, and architecturally feasible. MINERGIE® certified buildings are 60 percent more efficient in their heating systems than is required by current Swiss regulations. Also, the standard requires that construction costs not exceed standard costs by more than 10 percent, and in practice they tend to be only about two to four percent higher. Since 2000, the company has received MINERGIE® certification for 13 real estate projects with a total floor space of 151,000 square meters (1,625,350 square feet).

121. B2B—Business to business, as opposed to B2C—Business to customer.

122. Swiss Re. 2004. *Swiss Re Climate Specialist on TV*, http://www.swissre.com, viewed 1/29/06.

123. Swiss Re. 2003. *Becoming Carbon Neutral*, http://www.swissre.com, viewed 1/29/06.

124. Black powder is the oldest ballistic propellant for muzzleloaders and early cartridge arms composed of a mixture of potassium nitrate (saltpeter), charcoal and sulfur.

125. DuPont. 2006. *DuPont Heritage*, http://heritage.dupont.com/, viewed 1/8/06.

126. Solae is a manufacturer of soy protein and fiber ingredients in a joint venture with Bunge.

127. Pioneer Hi-bred International is a seed company specializing in biotechnology and genetic engineering.

128. Op. cit., Aston, A. and B. Helm. 2005.

129. Op. cit., Cogan, D. 2006.

130. DuPont. 2006. *Company at a Glance*, http://www2.dupont.com/Our_Company/en_US/glance/index.html, viewed1/8/06.

131. Ibid., DuPont. 2006. *Company at a Glance.*

132. DuPont. 2006. *Sustainable Growth.* http://www2.dupont.com/Our_Company/en_US/glance/sus_growth/sus_growth.html, viewed 1/21/06.

133. Op. cit., Aston, A. and B. Helm. 2005.

134. Thiemens, M. and W. Trogler. 1991. "Nylon Production: An Unknown Source of Atmospheric Nitrous Oxide." *Science*, 251(4996): 932-934.

135. The industry-wide agreement of N_2O producers included Asahi, BASF, Bayer, DuPont, ICI and Rhone-Poulenc.

136. HCFCs are generally considered interim replacements for CFCs. Their phase-out schedule is delayed compared to CFCs under the Montreal Protocol.

137. Warren, S. 2006. "DuPont Warns High Energy Costs will Hurt Profit", *The Wall Street Journal*, January 12: A6.

138. Reasons for shedding the nylon business were that the technology was "socialized" and it was not seen as a growth area for the company; it generated 25 percent of revenue but represented 40 percent of assets and was heavily dependent on fossil fuel.

139. Ranieri, J. 2005. *DuPont BioSciences: A Climate Change Best Business Practice*, Speech delivered to the California Public Utilities Commission, February 23, San Francisco, CA.

140. Tyvek® is a synthetic material made of high-density polyethylene fibers; the name is a registered trademark of the DuPont Company. It is a spunbonded olefin product that offers maximum protection and durability at a very light weight. For example, 100 10" x 12" envelopes weigh the same as 57 envelopes of the same size in 28 pound Kraft. Tyvek® is unaffected by moisture and inert to most chemicals. It has a number of uses, including siding for houses, envelopes, floppy disk and microfiche carriers where protection from acid, lint, and abrasions is needed.

141. DuPont. 2006. *Press Release: DuPont and BP Announce Partnership to Develop Advanced Biofuels.* (Wilmington: DuPont).

142. In some DuPont processes, steam is generated at a temperature above saturation (superheated). When process steps require saturated steam (which is cooler than superheated steam), water is sprayed into the superheated steam, cooling it down. This desuperheating water must be very high in quality so no deposits are formed when it vaporizes.

143. Speech delivered to the Clinton Global Initiative Panel on Climate Change, New York City, September 17, 2005.

144. Op. cit., Aston, A. and B. Helm. 2005.

145. Op. cit., Cogan, D. 2006.

146. For more on the Smart Trips program, see: http://www.smarttrips.org/, viewed 3/3/06.

147. For more on the One Ton Challenge, see: http://www.climatechange.gc.ca/onetonne/english/index.asp?pid=179, viewed 3/3/06.

148. Calculated for 2002.

149. Inskeep, S. 2005. "Gas Flaring Continues to Plague Nigeria." *National Public Radio*, Aug.25, http://www.npr.org/templates/story/story.php?storyId=4797953, viewed 10/18/05.

150. *National Public Radio* 2005. "Oil Firms Learn Trading Lessons." *National Public Radio*, May 9, http://www.environmental-finance.com/2003/0302feb/bpshell.htm, viewed 10/18/05.

151. Van der Veer, J. 2006. "A Vision for Meeting Energy Needs Beyond Oil." *Financial Times*, January 25: 21.

152. Reflects Whirlpool Corporation prior to the acquisition of Maytag Corporation in mid-2006.

153. The company has been broadly recognized for this commitment, including being named in 2005 as one of the 20 best corporate citizens by *Business Ethics Magazine.* In fact, the company has been named to the list every year since the magazine began publishing it six years ago.

154. *PRNewswire.* 2005. "Whirlpool Corp. to Cut Greenhouse Gas Emissions by 3 Percent From 1998 Levels." *PRNewswire,* http://web.lexis-nexis.com/5, viewed 9/7/06.

155. PA Consulting Group. 1992. *Ecolabelling Criteria for Washing Machines.* (London: PA Consulting Group).

156. *PR Newswire.* 2003. "Whirlpool Corporation Sues LG for Technology Patent Infringement." PR Newswire, http://www.whirlpoolcorp.com/news/release.asp?rid=221, viewed 10/28/05.

157. Horst, G. 2005. "Consumer 'White Goods' in Energy Management." http://ciee.ucop.edu/dretd/Whirlpool.pdf, viewed 10/28/05.

158. CO_2e.com. 2006. "What are Carbon Dioxide Equivalents (CO_2e)?" http://www.co2e.com/common/faq.asp?intPageElementID=30111&intCategoryID=93, viewed 1/24/06.

159. Greenhouse Gas Inventory Program. 2002. *Greenhouse Gases and Global Warming Potentials.* (Washington DC: U.S. Environmental Protection Agency): 9.

160. Energy Information Administration. 1997. *Mitigating Greenhouse Gas Emissions: Voluntary Reporting.* (Washington DC: U.S. Department of Energy).